Lecture Notes in Physics

Edited by H. Araki, Kyoto, J. Ehlers, München, K. Hepp, Zürich
R. Kippenhahn, München, D. Ruelle, Bures-sur-Yvette
H. A. Weidenmüller, Heidelberg, J. Wess, Karlsruhe and J. Zittartz, Köln
Managing Editor: W. Beiglböck

329

A. Heck F. Murtagh (Eds.)

Knowledge-Based Systems in Astronomy

A Topical Volume with Contributions by
A. Accomazzi H.-M. Adorf R. Albrecht A. Bijaoui
G. Bordogna R. Buccheri G. R. Cross V. Di Gesù
R. Gupta A. Heck H. Horstmann M. D. Johnston
M. J. Kurtz M. C. Maccarone G. Miller R. W. Miller
F. Murtagh P. Mussio R. Rampazzo A. Rampini
P. Schuecker D. Teuber M. Thonnat

Springer-Verlag
Berlin Heidelberg GmbH

Editors

A. Heck
Observatoire Astronomique
11, rue de l'Université, F-67000 Strasbourg, France

F. Murtagh
European Space Agency, Space Telescope – European Coordinating Facility
European Southern Observatory
Karl-Schwarzschild-Str. 2, D-8046 Garching, FRG

ISBN 978-3-662-13733-8   ISBN 978-3-540-46139-5 (eBook)
DOI 10.1007/978-3-540-46139-5

© Springer-Verlag Berlin Heidelberg 1989
Originally published by Springer-Verlag Berlin Heidelberg New York in 1989
Softcover reprint of the hardcover 1st edition 1989

2158/3140-543210

Table of Contents

New Directions

Glossary, Acronyms and Index

1 Foreword

With a nod towards Clausewitz' famous dictum on war and politics, one could say that applying artificial intelligence, expert systems, or knowledge based systems to astronomical problems is just programming by other means. But, as is amply exemplified in this book, this field offers substantial possibilities and a flexibility which classical programming does not.

The chapters of this book can be regarded as independent units. They offer a comprehensive synthesis of what is currently done in the general field of knowledge based systems by astronomers or by researchers linked with astronomical activities. Directions for future work are clearly pointed out and new orientations such as connectionism and neural networks are surveyed. Astronomical applications are stressed, rather than coverage being attempted of the entire artificial intelligence field.

There is already much interest in knowledge based system techniques in many branches of science. It has been quickly realized that increasing reliance on machines and powerful tools is the *sine qua non* for optimal management of data and exploitation of information contained in data. And, of course, both data and information are ever more rapidly accumulating and are ever more diverse.

The approach of astronomers to knowledge based disciplines appears to be somewhat slower than the approach of other natural scientists. We might indeed have to concur with what Mark Fox has said about software engineers: "The problem is not retraining [them] in AI; they learn quickly. The problem is *convincing* them that they need retraining".

On the other hand, it is interesting to note that, according to a recent study conducted by Edward Feigenbaum of Stanford University in California, by mid-1988 about 2000 expert systems had been put into productive use worldwide. Of these, 1500 were deployed in the USA, 250 in Japan and 250 in Europe. This represents only about 2% of expert system development shells that are thought to have been sold worldwide by over 100 vendors. Specialized expert systems built from scratch without the use of shells were not taken into account in the study. What this means is that the bulk of expert systems activity is concentrated in the stages of research and development. A significant increase in expert system use is thus foreseen for the 1990s.

The compilation of the papers presented in this volume has been a natural continuation for the editors of earlier work in bringing the astronomical community up-to-date with the most recent methodology at its disposal — as well as with corresponding applications already in use — in statistics, pattern recognition and related fields. Thus the interested reader can already refer to a textbook for astronomers on a selection of important methods in multivariate statistical analysis (Murtagh and Heck, 1987). A Working Group on *Modern Astronomical Methodology* regularly publishes a *Newsletter* (reprinted in the *Information Bulletin of the Strasbourg Data Centre*). A conference on *Astronomy from Large Databases: Scientific Objectives and Methodological Approaches* (Murtagh and Heck, 1988) dealt with ways to extract the "best" information from astronomical data.

A Technical Committee of the International Association for Pattern Recognition (TC 13, Astronomy and Astrophysics) concerns itself with this field, too, and the biennial colloquia on *Data Analysis in Astronomy* (proceedings published by Plenum Press, New York) held at the Ettore Majorana International Centre for Scientific Culture, Erice, Sicily, have become an established landmark on the astronomy/computer science interface.

It is a pleasure to thank all contributors for their prompt delivery of manuscripts which, together with the efficiency of the publisher, has made it possible for us to produce a volume with the timeliest and most up to date information available.

References

1. Murtagh, F. and Heck, A. (1987) *Multivariate Data Analysis* (Astrophysics and Space Science Library No. 131), Kluwer Academic Publishers, Dordrecht, xvi + 210p. (ISBN 90 277 2425 3, hardbound; ISBN 90 277 2426 1, paperback; ISBN 90 277 9154 6, floppy disk with source programs.)
2. Murtagh, F. and Heck, A. (1988) *Astronomy from Large Databases. Scientific Objectives and Methodological Approaches* (ESO Conferences and Workshop Procedings No. 28), European Southern Observatory, Garching bei München, xiv + 512p. (ISBN 3 923524 28 5.)

Telescope and Observatory Operations

2 Artificial Intelligence Applications for Hubble Space Telescope Operations

Glenn Miller
Astronomy Programs, Computer Sciences Corporation
Space Telescope Science Institute*
Homewood Campus
3700 San Martin Drive
Baltimore, MD 21218
USA

2.1 Introduction

A fundamental goal of observatory operations is to foster the scientific productivity and efficiency of the observatory and for many years computers have served an important role in observatory operations. Recent advances in the field of computer science known as artificial intelligence (AI) have allowed software to perform functions which previously required human attention. This paper examines the role of software based on AI techniques in observatory operations. Several examples of AI systems used in the science operations ground systems of the Hubble Space Telescope (HST) show how this technology has been successfully applied. The HST is an observatory of pioneering capabilities (see Hall, 1982, for reviews of the HST and its instruments). Orbiting at an altitude of 500 km, the HST will be above the bulk of the Earth's atmosphere, which will allow greater sensitivity, resolution and wavelength coverage than can be obtained from ground-based telescopes. A full complement of scientific instruments support imaging, spectroscopy, astrometry, photometry and polarimetry: the Wide Field/Planetary Camera (WF/PC), the Faint Object Camera (FOC), the Faint

---

* Operated by the Association of Universities for Research in Astronomy for the National Aeronautics and Space Administration.

Object Spectrograph (FOS), the High Resolution Spectrograph (HRS), the High Speed Photometer (HSP) and the Fine Guidance Sensors (FGS).

The scientific operations of the HST are conducted by the Space Telescope Science Institute (STScI). This includes the solicitation and selection of proposals from the worldwide astronomical community, scheduling the observing programs, executing observations, monitoring the spacecraft and instruments, and calibrating and archiving data. STScI staff work closely with mission operations engineers and scientists at NASA's Goddard Space Flight Center (GSFC).

The next section gives a brief sketch of HST science operations. Sections 2.3–2.9 describe several applications of AI techniques to HST operations. Given these examples, section 2.10 presents a general discussion of AI techniques and their role in developing science operations software. Section 2.11 speculates on future applications of AI technology to astronomical observatories. A final section summarizes the results.

2.2 Overview of HST Operations

In order to use the HST, an astronomer submits a scientific observing proposal to the STScI. Astronomers from a wide range of institutions and disciplines serve on the Time Allocation Committee (TAC) and advise the Director of the STScI in selecting proposals. Competition for HST time is expected to be keen; the oversubscription ratio (ratio of proposed to accepted observing time) may approach a factor of ten for the HST. Of the approximately 1000 proposals submitted yearly, only about 200–300 can be accepted for execution.

Following acceptance, the proposer must provide a more detailed description of the observing program, including the specification of each exposure, instrumental parameters, and the relationships between exposures (e.g. early acquisition, calibration, grouping of exposures, conditions for execution). The Proposal Entry Processor (Pep) System supports the staff of the STScI in receiving, selecting, evaluating and implementing the observing programs. Several of the AI applications described below are components of Pep, which is described more fully by Jackson et al. (1988).

A year's proposal pool will consist of perhaps 30,000 exposures on 3,000 celestial targets. Staff at the STScI produce a long term plan which spans the year's pool (see Johnston, 1989; Miller et al., 1988). Rather than allocating blocks of time to individual observers, observations with the HST will be scheduled at times which increase the overall efficiency of the observatory, so long as the constraints specified by the observer are satisfied. The long term plan must be updated throughout the year: occasionally, observations will not be executed successfully and the long term schedule must be revised accordingly. Targets of opportunity (e.g. supernovae) will also force revisions in the long term plan.

From the long term plan, small time segments (typically one week) are scheduled in detail, including instrument re-configurations, interruptions due to Earth occultation, spacecraft slews, etc. Given the detailed timeline, high level spacecraft instructions are attached to each activity, producing the Science Mission

Specification (SMS). The SMS is sent from the STScI to the Payload Operations Control Center (POCC) at GSFC where it is checked for errors or conditions which would affect the health or safety of the telescope or the instruments. From the SMS, the POCC prepares the actual binary command load for the two on-board computers controlling the HST. The SMS must be delivered to the POCC 30 days before the execution of the first observation on the SMS (therefore most observations are fully planned in advance).

All communications with the HST are via the Tracking and Data Relay Satellite System (TDRSS), which is shared among many satellites and the Shuttles. The POCC extracts TDRSS requests from the SMS, adds requests for command uplinks and tape recorder dumps, and forwards these requests to the TDRSS Network Control Center. Some requested links will not be available due to a higher priority user and this may require the STScI to reschedule some observations.

The HST is constantly monitored by the STScI and the POCC. About 20% of the observations will involve limited real time interaction and commanding of the HST, e.g. interactive target acquisition, selection of a particular spectral region in a variable object or choosing among pre-planned groups of observations. As data is returned to the ground system, it is inspected for transmission errors and other problems. In some cases, observations may have to be rescheduled. Data is processed through a series of routine calibrations at the STScI and archived. Observers may perform detailed analysis of the data at the STScI or at their home institution. After a year, the data is made available to the astronomical community via the HST archive, which is expected to be an important scientific resource.

This is just a brief sketch of what is a complex enterprise, involving the coordination of a large number of people and many different systems. In the next few sections, we describe several areas of HST operations which have seen the successful application of AI techniques. Several of these are used in routine operations — they are not just prototypes. The application of AI techniques to HST observation scheduling is so broad a subject that it is discussed in a separate chapter in this book (Johnston, 1989).

2.3 A Natural Language System for Proposal Selection Support

HST observing time is awarded via a peer-review process. The Time Allocation Committee carefully reviews each proposal and advises the Director of the STScI in the selection of the HST observing program. Several factors are considered in the selection of a proposal, including scientific merit and demand on the telescope resources. From one point of view, the selection process is essentially a "cost versus benefit" analysis where the benefit is related to the scientific merit and the cost is determined by the consumption of various telescope and ground system resources such as spacecraft time (exposure and overhead times), dark time, real time usage, parallel time, calibrations and data volume. The goal of this process is to select a set of proposals which maximize the scientific achievements of the

HST while not exceeding the limitations imposed by available resources. This is a difficult problem since a high oversubscription rate is anticipated.

The Time Allocation Committee Operations Support (TACOS) subsystem of Pep supports the selection process by providing a convenient and powerful way to track the various parameters in the selection process and the results of selection decisions. TACOS is described in detail in Hornick et al. (1987). An important consideration in the design of TACOS was the incorporation of a natural language interface to the users, the TAC members. Since the TAC members are donating a considerable amount of time in reviewing proposals, it would be unwarranted to burden them with learning a "computer" command language. An example (taken from Gerb, 1988) will illustrate this point. Instead of formulating a query in a relational database language such as IQL (Interactive Query Language, used for the Pep database):

```
range of a is vcoverpage
retrieve (a.id, a.dark_time, a.exposure)
where a.id >= 701 and a.id <= 707
```

the TACOS system accepts an equivalent query of the form:

```
for proposals 701 to 707,
display id, dark_time and exposure_time
```

A few additional examples will give a flavour for the power of TACOS. The specification of a set of proposals can be quite expressive:

```
display id, for proposals with time_on_target < 30,
                             except proposal 3
```

The user can define new words in the language:

```
define "summarize" as "display pi_last_name and title for"
summarize proposals 3-6
```

"PI" refers to Principal Investigator. TACOS can understand mathematical and logical operations on fields:

```
define "overhead_time" as "time - exposure_time"
define "small" as "with time_on_target <= 20"
define "medium-sized" as "with time_on_target > 20 and
                             time_on_target <= 100"
for all small proposals, display  overhead_time
for all medium-sized proposals
```

In the last line of this example, the command is omitted and only the proposal selection is specified. TACOS recalls the previous command and displays the "overhead_time".

Commands can be used to change the contents of the database. For example, the grade of a proposal can initially be set to be the average of the referee scores:

```
set grade to avg(scores) for all proposals
```

During the selection process, the TAC makes a number of decisions, such as assigning a proposal to a particular scheduling category or limiting the number of targets in a particular proposal:

```
for proposal 1175
set priority to 'medium'
limit number_of_targets to 15
```

To view the amount of resources allocated to accepted proposals, one could tell TACOS:

```
for all accepted proposals, display resources_used
```

(Definitions for "accepted" and "resources_used" are not shown for brevity.)

TACOS also provides a screen-oriented editor for displaying and modifying the database. Reports can be ordered by any field and cumulative totals of columns can be displayed. When TACOS cannot interpret a command, it explains the error and provides hints and suggestions to the user.

Conceptually, proposal data in TACOS is stored in a relational database consisting of a single relation. All information for a proposal is stored in a single tuple (row), keyed by proposal identification number. Each attribute (column or field) contains a quantity of interest such as number of targets, exposure time, data volume, instruments used, etc. TACOS provides a flexible, natural interface to query and modify the contents of this database. In this sense, TACOS is similar in purpose to a "spreadsheet" program, but supplying a much more powerful user interface.

An important consideration in the development of TACOS was flexibility in both the operations and maintenance of the system. Even though the basic procedures of the selection process are defined, a number of details are open to change (e.g. specific database fields, standard reports). TACOS meets the goals of flexibility in several ways: standard queries and reports are defined in an initialization file, which is easily modified. The ability to define new words (illustrated in the examples above) provides considerable flexibility during the TAC meetings. The grammar rules and keywords of TACOS are defined by input data files and are not embedded in the system's code. Likewise, the database fields and security levels are defined by the input database file. Not only does this allow new fields to be easily added to TACOS for proposal selection, it allow the system to be used on any single-table database.

TACOS is a successful implementation of natural language processing for a significant problem. However, some of the inherent limitations of TACOS should be noted. First, the domain of the problem is restricted to a single-table database and the permissible syntax is well defined. Additionally, TACOS need only parse single sentences, not groups of sentences or paragraphs. Once the sentence is successfully parsed, the semantic meaning (i.e. the code to execute the command) is known, since the grammar rules contain the code to be executed. Research on these broader issues is the subject of current AI research.

TACOS is implemented largely in Common Lisp, with some Digital Equipment Corporation (DEC) Vax utilities employed for displays and printing. These non-Lisp functions are isolated to allow porting to other computers and operating systems. TACOS uses a bottom-up shift reduce parser rather than a top down parser such as an Augmented Transition Network. A complete discussion of the design and implementation of TACOS is given in Hornick et al. (1987).

2.4 An Expert System to Transform a Scientific Program to Implementation Parameters

Before a proposal can be scheduled for execution by the HST, the information in the Proposal Forms must be translated into the parameters understood by the planning and scheduling software. The Pep Transformation system performs this conversion, and is the topic of this section. Further details are given in Rosenthal et al. (1986).

The HST Proposal Forms contain the description of the scientific observing program. The terms and logical structure of the forms are oriented towards the astronomer. In fact, astronomers can understand much of a proposal without the benefit of reading the proposal instructions. Of particular importance are the Target Lists and the Exposure Logsheet. The Target List provides information on the name, classification, position, and brightness of each target. The Exposure Logsheet lists each exposure and relationships between exposures such as: before, after, calibration for, acquisition for, sequential, conditional, select, etc. Special settings for various instrument parameters may be specified with optional parameters on the Exposure Logsheet. This information is stored in the Pep system. (Jackson et al., 1988, describes the Proposal Forms in more detail.)

Detailed, short term scheduling of the HST is performed by the Science Planning and Scheduling System (SPSS) developed by TRW, Inc. as part of the Science Operations Ground System (SOGS). The input to SPSS is quite different than the proposal form information. The SPSS input reflects the design of the scheduling software as well as spacecraft and orbital considerations. Input to SPSS is organized in a hierarchy as follows (described from the bottom up):

- An exposure is a single instrument operation. (Although the term "exposure" is used in the proposal forms, even this first level of the hierarchy cannot be simply mapped from Pep to SPSS.)
- An alignment is a set of exposures that can be taken without moving the telescope.
- An observation set is a set of alignments that can be performed using one set of guide stars.
- A scheduling unit is a set of observation sets and is the entity which is actually scheduled by SPSS. Scheduling units may be linked to each other.

The Transformation subsystem of Pep converts the Exposure Logsheet and Target List information to the SPSS hierarchy. This is not simply a matter of reformatting the information from one syntax to another: the Pep description is at a higher level while the SPSS description is much more exhaustive and detailed. There is not a direct one-to-one translation between the two.

At first it might seem that the problem was artificially induced by the proposal forms — if proposers submitted the information in the SPSS syntax, no translation step would be necessary. However this would put an unconscionable burden on the proposer for it would be necessary to understand details of how SPSS and the HST spacecraft work to describe even simple observations. The instruction manual would literally be thousands of pages in length (and therefore always out of date and largely ignored). The volume of information in a proposal would be unwieldy: since it is dealing at a lower level of detail, the SPSS description is roughly fifty times larger than a Pep proposal. Additionally, for most observations, there are several legal ways to transform from the Pep to SPSS representation and it would be difficult to ensure that similar observations would be transformed to similar SPSS representations.

The operations staff manually transforms proposals in a two stage process. First, an Operations Astronomer reads the Exposure Logsheet and Target Lists and generates a "script" which gives a skeleton of the hierarchical organization. The script lists the scheduling units, observation sets, alignments and exposures for the proposal and indicates where each Pep exposure falls in the hierarchy. An experienced astronomer can generate a few scripts per day. In the second stage, information is entered into SPSS by a Science Operations Specialist who uses the script as a guide. The Specialist is guided in this task by a book of written procedures which describes how various SPSS database fields should be populated on the basis of the Pep proposal entries. This stage typically takes a full working day or more, depending on the size of the proposal. Note that the Specialist is not merely entering data; a substantial knowledge of the transformation procedures is required.

Transformation by hand is an extremely labour intensive, tedious and error prone process. In addition, as experience was gained with the planning system, it was necessary to modify some of the transformation rules, which means that proposals needed to be transformed several times. All these factors established a clear need for an automated transformation system. In fact, these factors were recognized early in the development of STScI operations and were a major impetus for building the Pep system.

A rule based expert system was adopted for the core of Transformation for several reasons. Operations astronomers were willing to serve as experts for building the knowledge base. The solution process was well described by a series of loosely coupled if-then rules. Further, the experts were continuously revising their procedures to improve them and to reflect changes in SPSS, which was under development as well. It was felt that a rule based system would be more readily changed than a procedural program written in C or Fortran. DEC's OPS5 was chosen as the rule system.

A brief sketch of the Transformation system is as follows: a set of C routines retrieves proposal data from the Pep database and stores it in the appropriate OPS5 working memory elements. The OPS5 rules determine the SPSS hierarchy and the values of various scheduling parameters and write this information to an "assignment file". This file can be examined by an astronomer, and if needed, changes can be made using a text editor. Thus the system retains the speed of automated transformation and the flexibility of manual transformation for special case proposals. A procedural program reads the assignment file and enters this data into a copy of the SPSS database relations residing in Pep. Finally, the relevant records are extracted from the Pep system, transferred to the SOGS computers and loaded into the SPSS database.

A typical Transformation rule is:

```
(p find-parallel-with-mergeable-exposures
    (goal
            ^has-name           merge-exposures
            ^has-status         active
            ^task-list          find-potential-exposure-merges)
    (exposure-specification
            ^has-exposure-number                <primary-exposure>)
    (exposure-link
            ^is-linked-to                       <primary-exposure>
            ^has-link-type                      parallel_with
            ^has-exposure-number                <parallel-exposure>)
    (exposure-specification
            ^has-exposure-number                <parallel-exposure>)
    (mergeable-level
            ^symbol                             parallel-with
            ^value                              <parallel-with-level>)
    -->
    (make   mergeable-exposures
            ^first-exposure-number              <primary-exposure>
            ^second-proposal-id                 <parallel-proposal-id>
            ^second-version                     <parallel-version>
            ^second-exposure-number             <parallel-exposure>
            ^is-unmergeable                     false
            ^is-mergeable-level                 <parallel-with-level>
            ^merge-type                         parallel-with
            ^has-unique-label                   (genatom) ) )
```

The syntax of an OPS5 rule can be bewildering to the novice, but a few points will help clarify it. (Note that a number of other expert systems allow rules to be written in a natural language format.) The characters "(p" signify the beginning of a rule (rules are also called "productions") and are followed by the rule's name. Next come five condition clauses which are combined via a logical "and", that is, if the OPS5 working memory contains facts which match all of these conditions, then the rule is a candidate for execution. The "-->" identifies the end of the conditions and the beginning of the conclusions, that is, the actions to be taken if the rule is fired. This rule states that if the current goal is to merge exposures and there is a exposure which is linked to another exposure by a parallel-with special requirement, then the two exposures should be linked in a mergeable-exposures set with the indicated parameters.

The OPS5 rule base is goal oriented (which is essentially a way to impose a limited amount of procedural structure on the inherently opportunistic behaviour of the rules). The goal chain includes such tasks as reading input from the database, transforming target data, merging exposures into alignments, merging alignments into observation sets, merging observation sets into scheduling units and writing the assignment file. Some goals have sub-goals imposed as well.

The development of the Transformation system in many respects followed the textbook description of expert system development: a knowledge engineer (computer scientist) debriefed persons adept in manually transforming proposals and captured their knowledge in the form of production rules. The Transformation system has been used to support operations since December 1985, and currently the knowledge base contains about 850 rules.

So far this section has concentrated on the development of the Transformation system. An equally interesting topic is maintenance. The Transformation system has been continuously modified to reflect the changing requirements on the system and changes in SPSS itself. The system was well designed in that essentially all changes have been to the rule base and not other parts of the system. For the most part, the non-expert system components (programs which extract data from the Pep database, process an assignment file and extract data for the SPSS database) have been changed only to reflect changes in the definitions of the Pep or SPSS database relations. Even the impact of these changes was minimized by the use of "code writers" — database reports which produce fragments of code (such as C or IQL) from the relational database's data dictionary.

The most recent changes to Transformation have required iterative calculations as opposed to the loosely coupled prescriptions characteristic of the basic Transformation process. Iteration has been accommodated in the Transformation software in two ways: adding rules or calling out to a procedural language, performing the iteration and passing the results back into OPS5 working memory. Either method leads to code which is more difficult to understand and maintain than "pure" rules. The Transformation team is currently exploring ways to integrate procedural and rule based paradigms more easily than OPS5 allows. Another shortcoming of OPS5 which is becoming noticeable as the Transformation rule base grows is that condition clauses cannot be joined with a logical "or", e.g. a rule of the form "If A or B or C then..." must be coded in OPS5 as three separate rules. For rules with several or'd clauses, the number of combinations becomes large.

An informal, but important, way to gauge the maintainability of a system is how well it can be maintained once the original developer moves on to another project. Responsibility for Pep Transformation has changed three times in almost as many years. The new maintainers have quickly learned the rule base.

The implementation of OPS5 used for Transformation (from DEC) contains few development or maintenance tools. Lindenmayer et al. (1987) and Vick and Lindenmayer (1988) describe the development of a general methodology and tools for maintaining a rule base. These tools answer such questions as: "What rules are directly affected by rule A?", "Can the behavior of rule A ever affect rule B?", "Which rules generate conclusions that are never used by any other rule in the rule base?", "Which rules have conditions which can never be satisfied by any rule?" However, since these tools were developed recently, most of the maintenance of Transformation has been done without their benefit. A number of commercial expert systems offer some tools for debugging and maintaining a rule base.

2.5 An Expert System to Calculate HST Resource Usage

Throughout the process of selecting proposals and scheduling them for execution, it is important to have an estimate of the resources consumed by a proposal. Such resources include spacecraft time, dark time, data capacity, and communications contacts. For example spacecraft time includes not only the exposure time but time needed for moving the telescope, guide star acquisition, instrument configurations (e.g. moving a filter wheel), Earth occultation and data readout. Although estimates of the resources used by a proposal can be calculated from data on the proposals using simple formulae, an accurate calculation cannot be made until proposal exposures are mapped into the SPSS scheduling unit, observation set, alignment hierarchy since this breakdown determines the overhead of slews, instrument configurations, etc. (In fact, an exact calculation of resource usage cannot be made until an accurate predictive orbit is available and the timeline of observations is constructed.) This section briefly describes the Pep Resource Usage system. The reader should consult Jackson (1987) for details.

As noted in the last section, the rule based Pep Transformation system determines the mapping from Pep exposures to SPSS hierarchy. This lead to the development of a rule based Resource Usage calculator which is built as an extension of the Transformation system. Given the results of Transformation, the Resource Usage rules estimate slew times, guide star acquisition times, instrument configuration times, etc. to determine total spacecraft time, dark time, data volume, parallel observation time, communications usage and real time usage. An additional benefit of this approach is that as Transformation rules are modified, Resource Usage is automatically updated with no changes to the Resource Usage rules.

Perhaps the most interesting aspect of the Resource Usage expert system is that the Transformation rule base provided a natural foundation from which to build, even though Transformation was written without regard to the needs of resource usage. Only two changes were made to the Transformation system to accommodate Resource Usage. The file containing the Transformation rules was divided into two files, one containing the goals common to Transformation and Resource Usage and one containing the rules used only by Transformation (to generate the output). It was also necessary to add an additional data structure and rules to Transformation to do an accounting of parallel exposures.

2.6 An Expert System to Screen for Scientific Duplication

One question to be answered during proposal processing is "Does this exposure duplicate another exposure in the proposal pool or in the archives?". Checking for duplication is necessary to prevent waste of HST observing time and to ensure that observers do not duplicate observations which are reserved by the HST Guaranteed Time Observers.

The answer to this question depends on what is meant by "duplication". Certainly two exposures on the same target with the same instrument settings and exposure time are duplicates. Yet such a strict match of parameters is not always required. Different instrument settings or exposure times can produce observations that are essentially identical. Duplication can occur even when different instruments are used, since there is an overlap in the spectral, spatial and temporal capabilities of the instruments. The definition of duplication depends on so many variables, some quite subjective, that the final judgement must rest with an astronomer who understands the scientific goals of the proposals involved. Yet there are so many exposures (roughly 30,000 exposures per year and several years must be checked) that automated tools are required to check them all for duplication.

The Pep Duplication System (see Jackson, 1987, for details) is a tool which assists astronomers in identifying potential cases of scientific duplication. The role of the software is to identify cases where duplication is possible, assigning a confidence factor (high, medium or low) to the suspected duplication. The software must not be so simpleminded that it misses actual duplications nor can it be so undiscriminating that it identifies too many false candidates. Duplication candidates are identified by the following process: first, groups of targets which are near each other on the sky (typically within 0.1deg) are identified. Next, each target group is analyzed separately to find pairs of exposures which have similar instrument parameters. Finally each pair is examined to see if the targets are at statistically indistinguishable celestial positions or are separated by less than the chosen instrument aperture size. The first step is performed using a procedural program written in C, while the last two steps are performed by a rule based expert system written in OPS5.

The following is a typical duplication rule:

```
(p find-medium-very-similar-spectral-element-pairs
   (goal
        ^name              do-duplicate-checking
        ^condition         active )
   (very-similar-spectral-pairs
        ^uses-config-1                <config-1>
        ^uses-sp_1-1                  <sp_1-1>
        ^uses-config-2                <config-2>
        ^uses-sp_1-2                  <sp_1-2> )
   (exposure
        ^has-id                       <first-id>
        ^has-exp                      <first-exp>
        ^uses-config                  <config-1>
        ^uses-sp_1                    <sp_1-1> )
   (exposure
        ^has-id                       {<second-id> <> <first-id>}
        ^has-exp                      <second-exp>
        ^uses-config                  <config-2>
        ^uses-sp_1                    <sp_1-2> )
 - (potential-duplicates
                ^first-id             <second-id>
                ^first-exp            <second-exp>
                ^second-i             <first-id>
                ^second-ex            <first-exp>)
 -->
   (make        potential-duplicates
        ^first-id                     <first-id>
        ^first-exp                    <first-exp>
        ^second-id                    <second-id>
        ^second-exp                   <second-exp>
        ^confidence                   medium ) )
```

This rule states that if there are two very similar configuration-spectral element pairs, and one exposure uses one of the configuration-spectral element pairs and a different exposure uses the other of the pair and these exposures have not already been marked as potential duplicates, then mark them as potential duplicates with medium confidence. (The "–" preceding a clause indicates the logical "not" operation. Refer to section 2.4 for a brief introduction to the syntax of OPS5 rules.)

The Pep Duplication System only examines exposures of targets in the same group. An extension would be to consider targets of similar characteristics (e.g. quasars, or red giants) without regard to celestial location. To accomplish this, it would be necessary for the software to classify targets on the basis of the proposer's description and membership in standard catalogs.

As noted by Jackson (1987), the duplication rule base was written by an astronomer who had little experience with computer programming and no experience with AI techniques. After some initial consultation with a local OPS5 expert, the astronomer was able to successfully develop the rules without further assistance. This serves as a good example of an expert system built directly by the expert, without the aid of a "knowledge engineer". The non-procedural nature of rule based programming was quickly grasped by the astronomer and the peculiar syntax of OPS5 was quickly learned.

2.7 Expert Assistance in Preparing HST Observing Proposals

An astronomer desiring to use the HST must submit an observing proposal. As with any observatory, the proposer must understand the capabilities of the observatory and carefully complete the proposal forms. This is especially true in the case of the HST since the spacecraft activities are almost completely derived from information in the proposal (the observer does not control the telescope in real time). These and other factors have led to the development of the HST-Expert Assistant (HST-EA) for proposal preparation by staff at the Space Telescope — European Coordinating Facility (Adorf and di Serego Alighieri, 1989). The HST-EA provides four different functions: selection of an instrument configuration, exposure time calculation, a hypertext glossary of HST terms and a catalog of Guaranteed Time Observer observations. The HST-EA has been regularly used by European astronomers in the preparation of the first cycle of General Observer proposals.

In designing an observing program, the observer must select the appropriate instrument and various configuration parameters such as optical path, detector, and aperture based on the nature of the target and the type of measurement. Using the HST-EA, the user selects a category of observation (e.g. imaging, spectroscopy, astrometry) and target characteristics (e.g. point source) and is presented with the instrument configurations which are applicable for this observation. This function is of obvious benefit to the novice HST user, but can also assist more experienced users since there may be several different configurations, possibly using different instruments, which would provide the needed data. (This system is implemented using object oriented programming. A prototype of a similar system was implemented at the STScI using a rule based expert system.)

To use the Exposure Time Calculator, the user specifies a spectral type, apparent magnitude, background light level (one of high, medium or low) and the instrument configuration (currently the FOC, WF/PC, HRS and FOS are implemented; the HSP and FGS will be added in the future). The user can then input an exposure time and see the resulting signal-to-noise ratio, or, conversely, a signal-to-noise level can be specified and the required exposure time will be calculated.

The user interacts with the HST-EA via a mouse and menu interface and a high resolution graphics screen. For example, the Exposure Time Calculator displays signal-to-noise on a bar graph and the user can "click and drag" on the graph to change the value and see the effect on exposure time. This facilitates rapid experimentation to understand the effects of the target and instrument parameters.

Ordinary text is organized primarily in a sequential manner, with features such as a table of contents and index to allow non-sequential access. Hypertext systems greatly enhance the ability to establish and access cross references between items in the text (Walker, 1988). The HST-EA contains a hypertext glossary of definitions taken from the HST Proposal Instructions and Call for Proposals and is implemented using the Glossary utility on the Texas Instruments Explorer workstation. Within a definition of an item, words defined in the glossary are shown in bold type. Clicking on an item brings its definition into view. A user with a hypertext system can find information much more quickly than with paper documents or even conventional hierarchical computer help files. Such a system can be very useful to the novice trying to understand the many acronyms and terms used in a project such as the HST. Future work in this area could include distribution to the astronomical community of HST documentation in hypertext format on popular microcomputers.

The HST-EA instrument configuration selector and exposure time calculator are implemented in Common Lisp and the Knowledge Engineering Environment (KEE) expert system shell. The facts in the knowledge base are stored using KEE's object oriented programming system. Data is grouped into classes, classes of classes, etc. For example, the class of FOC filters contains classes of neutral density and passband filters which in turn are comprised of individual filters. Since the basic instrument data was available as tables in the MIDAS (Munich Image Data Analysis System) image processing system, the HST-EA includes a component to convert these tables to KEE knowledge base format. As the instrument data is updated (e.g. improved measurements of instrumental efficiencies), the knowledge base is easily updated.

2.8 An Expert Assistant for Astronomical Data Analysis

In order to be useful, raw data from an instrument must be calibrated: the instrumental signature must be removed and data must be converted to physical units (e.g. counts to magnitudes). This section describes Johnston's (1987) prototype expert system for calibrating data (see also Thonnat and Clément, 1989).

Observational astronomers are acquainted with data reduction and analysis systems, such as Space Telescope Science Data Analysis System (STSDAS), Munich Interactive Data Analysis System (MIDAS) and Image Reduction and Analysis Facility (IRAF). These systems provide a variety of analysis functions (such as subtraction of flat fields, averaging, bad point rejection, etc.) and the user must assemble the functions in a particular order for the task at hand (wavelength-calibrated spectrum, surface photometry, etc.). The great power and flexibility of these systems can also present problems for the user: learning a system can be hard for a novice, and even experienced users may be unfamiliar with large parts of the system. Traditionally, users have relied on the goodwill of local experts and, to a lesser extent, user manuals and help files. In addition, generating the necessary script file of commands to reduce an observing run can involve a substantial amount of work by the user.

An alternative is provided by Johnston's expert data analysis assistant. Given a description of the data from the user, the system uses its knowledge of calibration to generate a generic plan. It then translates the plan into a specific data reduction language, in this case, either STSDAS or MIDAS. The prototype was restricted to calibration of CCD images but was designed to be extensible: augmenting the system to handle other calibration tasks does not require changes to the reasoning mechanism, just the addition of facts defining new instrument modes, new classes of data (e.g. spectra), and new analysis tasks (e.g. wavelength calibration of spectra). Tasks are hierarchically composed of tasks and at the bottom of the hierarchy are primitive tasks which can be directly accomplished via STDAS or MIDAS commands.

Thonnat and Clément (1989) describe a similar system which has been applied to the problem of processing images of galaxies. Using knowledge of a library of image analysis functions, the system plans and executes the necessary commands to detect a galaxy and calculate parameters such as size, ellipticity and luminosity profile. Interestingly, the results of this system are used by another expert system (Thonnat, 1988) which then classifies the galaxy according to the revised Hubble scheme.

2.9 Monitoring the HST

The HST will be monitored by engineers and astronomers at GSFC and STScI in order to ensure the health of the HST and the quality of the scientific data. When a TDRSS communications downlink is available, the engineering telemetry is monitored continuously in real time, otherwise it is stored on an onboard tape recorder for later playback to the ground. The spacecraft is continuously gathering information about the status of its systems — several thousand measurements are made of voltages, temperatures, relay settings, position encoders, etc. These are polled at least every two minutes (some as often as 40 times per second) and transmitted to the ground via the telemetry stream. In addition, the science data is an important source of information on the behaviour of the instruments (e.g. changes in dark count, wavelength scale, geometric distortions, noise).

Monitoring is an important task primarily because the cost of problems can be quite high: loss of important observations or even damage to the spacecraft or instruments. Successful monitoring will allow operators in some cases to detect and correct a problem before it becomes serious, and in situations where a problem cannot be avoided (e.g. sudden failure of a component), to understand the problem and devise a recovery.

Monitoring the thousands of HST telemetry parameters presents a challenge to operations staff and the supporting software. Traditionally, software has performed rather simple limit checking on parameters and humans have performed the bulk of the analysis from video displays of telemetry information. Surprisingly, the displays in some spacecraft control centres have not been well-designed for human understanding. They can be densely packed, labelled with a cryptic code indicating what is being monitored (e.g. "QDFHILO" for a particular relay), and the value displayed in raw sensor units instead of physical units (such as degrees Centigrade). Staff members may have to monitor many displays. Since the telemetry data are measurements of individual sensors, it is often the case that a meaningful state of the telescope or instrument is a function of several readings, e.g. an instrument is in the operate state if relay X1 is closed and encoder Y7 has a reading of 3 or if relay X2 is open and encoder Y9 has a reading of 4. These factors can lead to a situation where monitoring is tedious and prone to human error. Additionally, it requires many months to train a new staff member or to retrain staff as new monitoring procedures are developed.

One major improvement in such a system is to display the information in a more readable style, e.g. using graphics displays to show a schematic of the spacecraft and location of the sensors, color to indicate certain conditions, etc. The HST Observatory Monitoring System (OMS: Sen, 1988) has been developed by the STScI to monitor HST telemetry and effectively present the information to the operators. The University of Colorado's Operations and Science Instrument System (OASIS: Jouchoux et al., 1987) is another example of a monitoring and command system incorporating modern display techniques. (Since this paper is concerned with AI software engineering techniques, it is important to observe that good software has been written without explicit use of AI languages: OMS is written in Fortran and OASIS is written in Ada.)

Another way to improve spacecraft monitoring is to increase the sophistication of the software beyond simple limit checking. For example, OMS and OASIS can convert telemetry into physical units, test combinations of monitors and do some trend analysis. A knowledge based monitoring system for the HST Pointing Control System and Flight Software was developed by Laffey et al. (1988b) using the L*STAR expert monitoring system (Lockheed Satellite Telemetry Analysis in Real Time). The system has been used to monitor the HST during ground tests. L*STAR consists of three independent processes: data management, inference and user interface. The data management process receives the incoming telemetry and routes sets of data to the other two processes. The inference process analyzes the telemetry. The user interface process graphically displays information from the other two and allows the operator to enter commands.

The inference process uses a rule based expert system and procedural code to analyze the data. Analyses can be invoked in three ways: at fixed intervals of time, when data changes and when needed to achieve a goal. A typical rule can be paraphrased "If the pointing control system is in manoeuvre mode, and the value of monitor QDSTDCP has been decreasing for the last ten seconds, and the value of monitor QDSTCP > 0.000043 and the value of monitor QDFHILO =1, then send an alert that the rate gyro assembly is not in high mode and set its status to abnormal".

An advantage of a rule based expert system in monitoring is that the rules can be written in isolation from other rules in the system. It is the responsibility of the inference engine to execute the rules under the correct conditions. In procedural code (i.e. a series of if-then or case statements in C or Fortran) the programmer must ensure that the code will be executed at the correct time, i.e. that control of the program will pass through those statements. In a system consisting of many hundreds of conditional statements, this is impossible for all practical purposes.

The inference process initially tells the data management process what monitors to send, how to smooth data, and the minimum amount a monitor must change before it is reported. The inference process may later request the data management process to change this reporting. This allows the inferencing to focus on the appropriate information for a certain condition and not be overwhelmed by attempting to examine too much information.

It is not sufficient to ensure that the spacecraft and its systems are in a legal state — it is important to verify that they are in the correct state, for example that the correct instrument is in operating mode, the correct filters are in position, or that guide stars are being acquired. Since the HST is largely preplanned, this equates to comparing the current state of the spacecraft with the "Mission Schedule" which is a detailed list of spacecraft activities generated when the spacecraft computer command loads are generated. (The HST OMS does perform this comparison.) Spacecraft with more autonomous behaviour would present more difficult diagnosis problems.

An interesting review of real time expert systems in a variety of fields is given in Laffey et al. (1988a). A number of general purpose systems for real time applications are reviewed. They also discuss theoretical considerations for real time expert systems, including the issue of guaranteeing response times. They note the performance requirements that real time monitoring places on the expert system — the system should draw conclusions from the incoming data on the same timescale as changes in the data occur. In observatory monitoring, the requirement for real time performance is not quite so strict since the expert system works in conjunction with humans and procedural software. Fast procedural software can assure that critical values are always checked by simple tests, while humans are available to assess situations which are beyond the expert system. The requirement on the expert system is to notify the user when the inferencing process falls behind the incoming data.

2.10 The Role of Artificial Intelligence Tools in Observatory Operations

The previous sections have given an empirical introduction to AI systems used in observatory operations. This section takes a more general look at AI.

2.10.1 What is Artificial Intelligence?

The search for artificial intelligence (AI) is a fascinating scientific problem, with ultimate goals of understanding human intelligence and creating intelligent machines. How can we know when a machine is intelligent? The pioneering computer scientist, Alan Turing (1963), put forth the following test. Imagine three isolated rooms where the only communication possible is a teletype link. Typing on a keyboard prints the message in the other rooms. One room contains a human interrogator. The other two rooms contain one subject apiece, a human and a machine. The interrogator must determine which is which. If the machine can convince the interrogator that it is the human, then it has displayed intelligence. Turing envisioned that the machine would be capable of quite general and sophisticated behaviour, for example, if asked to divide two 20 digit numbers, the machine might pause for a few minutes and then give a plausibly wrong answer.

Although no machine has passed this test, this does not mean that no progress has been made towards the goals of AI. To better understand this progress, it is useful to adopt the pragmatic definition offered by Rich (1983): "Artificial intelligence is the study of how to make computers do things at which, at the moment, people are better." An interesting consequence of this definition is that the realm of AI problems is ever shrinking, for once a problem is solved, it becomes mundane and is no longer AI.

In the spirit of Rich's definition, this paper has addressed the question "How can AI techniques be used to automate tasks in observatory operations which previously required humans?" The experience with HST science operations is that AI tools have made it possible to develop a number of useful systems that could not have been built using more classical software techniques.

This section presents an overview of some important topics in AI and tools for development of AI programs which have direct application to observatory operations. This presentation is necessarily selective and some important fields in AI are not mentioned at all. A brief introduction to AI is given by Schank (1987) and more thorough treatments are found in books such as Rich (1983), Charniak and McDermott (1987) or Winston and Horn (1988). An interesting account of the history of AI research and the relation between human and artificial intelligence is given in Dehn and Schank (1982).

2.10.2 Fields of AI

Several fields of artificial intelligence have applications to observatory operations:

Learning — Whatever intelligence is, a key component of intelligence must be learning: the ability to extract knowledge from experiences and apply it to new situations. By this standard, the intelligence of many expert systems can be questioned, for although they can perform certain tasks as well or better than a human, they cannot perform even closely related tasks. Several programs have exhibited learning, and perhaps the best examples are Lenat's (1983; see also Rich, 1983) programs, which, given a few basic concepts of set theory, proceeded to discover a number of interesting results in set theory. An example of a learning system in a laboratory setting is found in Buchanan et al. (1988). Using test data generated from a particle accelerator simulation program, an expert system induced general rules for diagnosing and correcting misalignments in the particle beam. These general rules were then used by another expert system to successfully diagnose beam alignment problems. Thus the learning system served as the knowledge engineer. (An analogous problem in astronomy is the alignment of optical components.)

Search — Many problems can be characterized as searching for one or more solutions from many possible alternatives. Examples include scheduling, games (e.g. chess), a robot planning a series of tasks, and speech recognition. In principle, search problems can be solved by trying all possibilities and selecting the best. In practice, the time it takes to do this may grow exponentially with the size of the problem and is therefore useless for problems of even modest size. For example, if it takes one week to generate a week long schedule for the HST by exhaustive examination of all possibilities, generation of a month's schedule will take well over a year and the time to generate a 6 month schedule will approach the age of the Universe. This so called "combinatorial explosion" is well known to anyone who has had to solve real search problems. Garey and Johnson (1979) give an understandable yet mathematically sound discussion of the general class of computationally hard problems. Techniques from the field of operations research (e.g. linear programming, see, for example, Hillier and Lieberman, 1986) are applicable to many search problems, but in making the problems computationally tractable, they introduce simplifying assumptions which can take the problem too far from reality. AI research has resulted in a number of powerful methods for attacking complex search problems (see Pearl, 1984) and these find a variety of applications in observatory operations.

Reasoning — The ability to store a large amount of knowledge in a useful way and reason with the knowledge has been one of the most notable achievements of AI research. Non-AI programs may also contain a large amount of information, but usually store it in such a way that it is extremely difficult to apply this information to related problems. The fundamental characteristics of AI systems in this respect is the *explicit* declaration of knowledge (not implicitly contained in the computer code) and the explicit declaration of characteristics of the knowledge ("meta-knowledge"). Expert systems serve as a good example. Systems have been developed which can reason with uncertainty, make inferences from related information, and which can explain how conclusions were drawn from the knowledge in the system.

Natural Language — Communicating with computers via a natural language (e.g. English) instead of an artificial language (e.g. Fortran) is the goal of natural language systems. Commercial systems for understanding natural language text are available. Computer synthesized speech is in everyday use (e.g. banking machines and toys). Speech recognition by computers is beginning to be a practical option for real systems, but only if the computer is trained by each speaker in advance and the vocabulary of discourse is limited (which are reasonable restrictions for operational applications).

2.10.3 AI Tools and Techniques

Research in AI has resulted in a number of powerful tools for software development, including languages, methods for representing and reasoning with knowledge, and programming environments. These topics are briefly reviewed in this section (see also Tichy, 1987).

The Lisp programming language has played an important role in artificial intelligence. Just as Fortran can be called the programming language for physical science, Lisp can be called the programming language of AI. (Lisp, like Fortran, is one of the earliest high-level computing languages, having been developed in the 1950's. "Lisp" stands for List Processing, since the principal data structure is the list.) Lisp provides a number of features which are very useful for symbolic computation, including rich data structures and powerful program control methods. Lisp implementations include an interpreter which executes code incrementally (like the Basic programming language). This provides the developer with an interactive view into the program and speeds development and debugging. Lisp is a flexible language: it is easy to add definitions which extend the language and it is straightforward to implement programs which write new Lisp code and then execute it. Tatar (1987) provides a good introduction to Lisp for a programmer proficient in other languages. Winston and Horn (1988) provides an introduction to Lisp with many interesting projects and examples.

Lisp has overcome weaknesses in two areas vital to practical applications: speed and standardization. In the past, Lisp implementations offered only an interpreter, so Lisp programs were slow compared to compiled programs. Current Lisp compilers produce code as fast as that of Fortran or C compilers. At one time there existed a number of incompatible dialects of Lisp. Changes within a given dialect sometimes hampered the maintenance of a program. Common Lisp (Steele, 1984) is now the accepted standard for the language and is implemented on a wide variety of machines, from microcomputers to mainframes. Practical applications often require interfacing diverse software and hardware systems, for example, querying a database or calling from one language a software library written in another language. Lisp programs are no more difficult to interface than other languages.

Object oriented programming has proven to be an important way to represent and manipulate knowledge. It allows the developer to represent complex relationships between objects in a system and encourages a separation of the interface to a software module from its implementation details. Object oriented programming allows related objects to share data values and code. This allows one piece of code to apply to many different (but related) objects and eliminates the need seen in other systems to write new code for each new variation of an object. The Common Lisp Object System (CLOS) is emerging as the standard for Lisp systems. Keene (1989) provides a good introduction to object oriented programming in CLOS.

A large number of systems (also called "shells") for developing AI applications are available. Typically these systems are built from Lisp although some are implemented in C or other languages. These include Knowledge Engineering Environment from IntelliCorp, Automated Reasoning Tool from Inference, KnowledgeCraft from Carnegie Group, and OPS5 from DEC. Representative tools include rule based inferencing, object oriented programming, natural language processing, hypothetical reasoning and neural networks. By using these shells, a developer builds on the work of others and avoids having to write yet another rule interpreter, etc. Disadvantages of the shells can include the slower execution speed, cost of the software, limitations in the scope of the tools, and the problems inherent with relatively new software packages. A review of AI tools is given in Gevarter (1987). Many shells are available on microcomputers such as the Macintosh.

Equally important as AI languages are the tools that have been closely associated with them. These include large, high resolution graphics screens, window and mouse oriented user interface, and integrated software development tools such as a debugging aids, ability to browse data structures in memory, ability to compile changes to code, and cross-references of software dependencies.

One of the results of the languages and tools of AI is the ability to quickly implement software so that ideas can be tried and modified. This is particularly important for problems where the requirements are poorly defined or too complex to be specified fully in advance. The advantages of rapid prototyping have been demonstrated in a variety of projects (Agresti, 1986).

This section has tried to sketch some of the areas of AI research which can help observatory operations and to mention the features of AI tools which make them suitable for software development. However, this enthusiasm should be put in perspective. It is not meant to imply that only Lisp environments provide these tools nor that AI tools can be developed only in Lisp environments. Analogs to the tools described above can be found in many Fortran or C programming environments (often developed by the programmers if they are absent from the commercial software). The point to be understood is that AI tools should be considered as candidates for implementing operational systems; the builders of the system should choose the right programming tool for the job at hand. AI tools do not automatically result in well designed, useful software — it is possible to produce poor systems in any language.

2.11 Future Applications for HST and Other Observatories

In the preceding sections we have sought to show the power of AI technology and that useful, practical systems can be built from this technology. In this section we speculate on how AI techniques can be extended to more complex problems in operating the HST and other observatories. From a different perspective, Smith (1986) gives an interesting account of the history and future of computer applications for data collection and analysis.

Proposal Preparation: Providing the community with better access to expert advice on the telescope and instruments is an area where expert systems can be quite helpful as shown by the Proposal Preparation Expert Assistant (section 2.7).

The HST Proposal Forms allow a proposer to specify exposures and their interrelationships in a flexible manner, using terms which are familiar to astronomers. However the Proposal Forms have a strict syntax which must be obeyed and are far from a natural language description. It would be useful to allow proposers to specify a more natural description of the program, e.g. "Observe the core of each globular cluster using the WFC U, B, V filters. Take two exposures on each object. They need not be consecutive, but the telescope roll angle should be the same for both exposures. Depending on the results of the WFC exposures, followup observations will be made with the FOC...". From this description a natural language expert system could complete the HST Proposal Forms, specifying detailed instrument parameters and linkages between exposures. Such a system could have in its knowledge base a number of typical observing scenarios and choose ones most appropriate for the observations at hand. Understanding of the scientific requirements such as spatial and spectral resolution would allow instrument settings and calibrations to be specified.

Of course the very richness and flexibility of natural language allows uncertainty and ambiguity into a description (leading to the familiar adage "Don't do what I say, do what I mean."). The proposer would need to review the system's output to be sure that the intent was properly understood. To reduce the possibility of misunderstanding, the system would need to inform the user of what ambiguities were encountered and how they were resolved, and what assumptions were made (e.g. that the field of view is sufficiently simple that an onboard acquisition can be used instead of an ground-assisted interactive acquisition).

The need for an explicit description of an observing program is not unique to the HST. Many observatories are experimenting with the concept of "service" or "absentee" observing, where the observatory staff collects data for the proposer as conditions permit. Likewise, automated ground telescopes and spacecraft need an observing plan. A clear and effective means of communicating the science requirements is important. Such a description should not force the proposer to commit to actions prematurely and allow sufficient flexibility to defer certain decisions until execution time (see also Jackson et al., 1988).

Proposal Selection: The TACOS system merely tracks variables in the decision process; it does not suggest a selection of proposals according to cost and benefit parameters. A possible extension in capabilities would be to supply a reasoning mechanism which, given measures of scientific merit from the TAC members, suggests possible combinations of selected proposals and the implications of the selection. Since there are multiple criteria to be optimized (scientific merit, resource consumption, balance among scientific disciplines, balance between large, medium and small proposals, etc.) and several possible solutions, this is not a simple problem of allocating resources to the highest ranked proposals until one or more resources are exhausted. It is necessary to search for all combinations of proposals which, considering the uncertainties in the parameters, are close to optimal.

To illustrate with an example, the system might find one combination of proposals which meets resource limitations on exposure time but devotes less time to large programs than initially desired. A competing selection scenario might meet most of the resource limitations but exceed one resource limitation (perhaps real time usage) by a small amount. A third possibility might have the desired balance among scientific disciplines but a slightly lower overall scientific merit. Human judgement is required to pick among these competing scenarios. It is recognized that such a system might be quite controversial if it were perceived that humans were abrogating responsibility to a machine. However, properly implemented and used, such a tool could allow the TAC members to consider a wider range of possibilities while retaining control over the process. At this time, it is not clear if the search for good proposal selections is sufficiently hard to require such automated help.

Observation Monitoring: Section 2.9 described a system used to monitor engineering telemetry from the HST. Similar systems are being used for other spacecraft (see papers in Rash and Hughes, 1988) and should play an increasingly important role for space observatories and large ground-based observatories. Monitoring can be divided into three tasks: detection of a problem, diagnosis of causes, and recovery from the problem. Current systems have focused mostly on detection with some abilities for diagnosis and recovery; future systems should provide capabilities in all areas. Software which can understand human speech would be very useful in an operations control centre.

As such systems mature, we could expect them to incorporate AI learning and discovery techniques, that is, to detect a new problem state on the basis of similarity to a previous problem or because the system was exhibiting an interesting behaviour.

In addition to monitoring engineering telemetry, the science data are also a valuable source of information: if the engineering telemetry is monitored to ensure the health and safety of the spacecraft, the goal of monitoring the science data is to ensure the scientific quality of the observations. Changes in such items as the instrumental dark count, geometric distortions, or wavelength scale may indicate the onset of a problem. In fact, such changes in the science data may indicate problems far in advance of engineering telemetry. Since the changes are likely to be more subtle, they would present a greater challenge to a system.

Data Traceability: The process of constructing and executing an observing program involves several stages, such as planning the observations, devising calibration observations, scheduling the telescope, selecting instrument commands, and collecting and reducing the data. In each of these stages the observations are represented in a different way, tailored to a particular perspective. For example, the list of HST spacecraft activities and commands (the Science Mission Specification or SMS) is vastly different in structure and content from the Proposal Forms, although they both embody the same scientific objectives. This transformation of observations from one representation to another is an intrinsic part of observing as one proceeds from a high-level goal (e.g. determine the lithium abundances in a sample of red giant stars) to a specific realization of that goal (point at HD 5234, position the grating, open the shutter, etc.).

Traceability is defined as the ability to track each element of an observing program through the different representations and is essential to the routine operations of an observatory. Typical traceability issues include: What data files comprise a particular observation? What calibrations should be used? Was an observation successfully executed? An error is discovered in the target coordinates, what information in what systems needs to be corrected? Traceability can be complicated by changes made during the process (e.g. rescheduling failed observations) and the fact that the different systems involved may be staffed by different organizations at different geographical locations. Miller et al. (1988) have examined the problem of traceability in operations and in data archive construction, and present a model for developing a system to answer traceability questions.

Archiving and Analysis: The HST will produce terabytes of data. After a proprietary period, HST observations will be made public. Storage of this amount of information has required new technologies such as optical disks and fast relational databases (such as the HST Data Management Facility, see Richmond et al., 1988). The next challenge facing developers and users of archive systems is how to find the needed information among all those many bytes. Traditional data retrieval systems have required the user to know a good deal about how the information is stored (e.g. by target, by observation number, by right ascension) and often the user has had to learn a specialized query language. Current progress in AI technology (see Rosenthal, 1989, and Adorf et al., 1988, for a review) makes it reasonable to foresee systems where the user can receive answers to questions such as: "Do any other high redshift quasars have a CIII line that looks like this one?" "How many other HD stars have been observed in this spectral region?".

Several other chapters in this volume show how knowledge based systems can be used in data analysis.

2.12 Summary

Using Hubble Space Telescope operations as an example, this paper has shown practical applications of AI techniques to observatory operations including proposal preparation, proposal selection, proposal transformation, resource usage, duplication, observation monitoring and data analysis. Several of the systems are in routine use by operations staff and solve problems which formerly required highly trained human experts. The power of AI techniques results from several factors including sophisticated development tools, powerful ways to represent and reason with knowledge, and an expressive user interface.

Although this paper has used the HST as a case study, most features of HST operations are common to other observatories, both space- and ground-based. NASA's great observatories such as the Advanced X-Ray Astrophysics Facility (AXAF), the Space Station, and ground-based telescopes such as the European Very Large Telescope and Texas-Penn State Spectrocopic Survey Telescope can benefit from applications of AI technology.

Much of the work described here was part of the development of the Proposal Entry Processor (Pep) System at the STScI. It is a pleasure to acknowledge the contributions of the Pep team: William Cohen, Andy Gerb, Robert Jackson, Mark Johnston, Kelly Lindenmayer, Robin Lerner, Patricia Monger, Don Rosenthal, Jim Sims, Lyle Sutton, and Shon Vick. The HST proposal preparation Expert Assistant was developed at the ST European Coordinating Facility by Hans-Martin Adorf and Sperello di Serego Alighieri. Amit Sen provided information about the HST Observatory Monitoring System and real time monitoring problems. Special recognition is due Mark Johnston, who has been a guiding force behind science operations software at the STScI. I thank Hans-Martin Adorf, Robert Jackson, Mark Johnston, Kelly Lindenmayer, Jeff Sponsler, and Shon Vick for discussions and comments on the manuscript.

References

1. Adorf, H.-M., Albrecht, R., Johnston, M. and Rampazzo, R. (1988), "Towards heterogeneous distributed very large databases", *Astronomy From Large Databases: Scientific Objectives and Methodological Approaches*, F. Murtagh and A. Heck, (eds.), European Southern Observatory, Garching bei München, 137–142.

2. Adorf, H.-M., and di Serego Alighieri, S. (1989), "An expert assistant supporting Hubble Space Telescope proposal preparation", *Data Analysis in Astronomy III*, V. Di Gesù, L. Scarsi, P. Crane, J. Friedman, S. Levialdi and M.C. Maccarone (eds.), Plenum Press, New York, in press.

3. Agresti, W. (1986), *New Paradigms for Software Development*, IEEE Computer Society Press, Washington, D.C.

4. Buchanan, B., Sullivan, J., Cheng, T. and Clearwater, S. (1988), "Simulation-assisted inductive learning", *Proceedings of the Seventh National Conference on Artificial Intelligence*, Morgan Kaufmann, San Mateo, CA.

5. Charniak, E., and McDermott, D. (1987), *Introduction to Artificial Intelligence*, Addison-Wesley, Reading, MA.

6. Dehn, N. and Schank, R. (1982), "Artificial and human intelligence", *Handbook of Human Intelligence*, R. Sternberg (ed.), Cambridge University Press, Cambridge, UK, 352–391.

7. Garey, M. and Johnson, D. (1979), *Computers and Intractability: A Guide to the Theory of NP-Completeness*, Freeman, New York.

8. Gerb, A. (1988), "An overview of the TACOS system", unpublished lecture notes.

9. Gevarter, W. (1987), "The nature and evaluation of commercial expert system building tools", *Computer*, **23**, 24.

10. Hall, D., ed. (1982), "The Space Telescope Observatory", NASA CP- 2244, NASA, Washington, D.C.

11. Hillier, F. and Lieberman, G. (1986), *Introduction to Operations Research*, 4th ed., Holden-Day, Oakland, CA.

12. Hornick, T., Cohen, W. and Miller, G. (1987), "A natural language query system for Hubble Space Telescope proposal selection", *Proceedings of the 1987 Goddard Conference on Space Applications of Artificial Intelligence and Robotics*, NASA, Greenbelt, MD.

13. Jackson, R., Johnston, M., Miller, G., Lindenmayer, K., Monger, P., Vick, S., Lerner, R. and Richon, J. (1988), "The Proposal Entry Processor: telescience applications for Hubble Space Telescope operations", *Proceedings of the 1988 Goddard Conference on Space Applications of Artificial Intelligence*, J. Rash and P. Hughes (eds.), NASA Conference Publication 3009, NASA, Greenbelt, MD, 107–124.

14. Johnston, M. (1987), "An expert system approach to astronomical data analysis", *Proceedings of the 1987 Goddard Conference on Space Applications of Artificial Intelligence and Robotics*, NASA, Greenbelt, MD.

15. Johnston, M. (1989), "Knowledge based telescope scheduling", this volume.

16. Jouchoux, A., Cowley, J., Davis, R., Hansen, E., Klemp, M., Lasater, S., Mullens, D., Sparn, T. and Tate, G. (1987), "Developing a spacecraft monitor and control system in Ada", paper presented at the Joint Conference Fourth Washington Ada Symposium Fifth National Conference on Ada Technology.

17. Keene, S. (1989), *Object Oriented Programming in Common Lisp*, Addison-Wesley, Reading, MA.

18. Laffey, T., Weitzenkamp, S., Read, J., Kao, S. and Schmidt, J. (1988a), "Intelligent real-time monitoring", *Proceedings of the Seventh National Conference on Artificial Intelligence*, Morgan Kaufmann, San Mateo, CA.

19. Laffey, T., Cox, P., Schmidt, J., Kao, S. and Read, J. (1988b), "Real-time knowledge based systems", *AI Magazine*, 9, 27–45.

20. Lenat, D. (1983), "Eurisko: a program that learns new heuristics and domain concepts. The nature of heuristics III: program design and results", *Artificial Intelligence*, March.

21. Lindenmayer, K., Vick, S. and Rosenthal, D. (1987), "Maintaining an expert system for the Hubble Space Telescope ground support", *Proceedings of the 1987 Goddard Conference on Space Applications of Artificial Intelligence and Robotics*, NASA, Greenbelt, MD.

22. Miller, G., Johnston, M., Vick, S., Sponsler, J., and Lindenmayer, K. (1988), "Knowledge based tools for Hubble Space Telescope planning and scheduling: constraints and strategies", *Proceedings of the 1988 Goddard Conference on Space Applications of Artificial Intelligence*, J. Rash and P. Hughes (eds.), NASA Conference Publication 3009, NASA, Greenbelt MD, 107–124; reprinted in *Telematics and Informatics* (1987), 4, 301.

23. Miller, G., Monger, P. and Johnston, M. (1988), "Data traceability in the construction of astronomical archives", *Astronomy From Large Databases: Scientific Objectives and Methodological Approaches*, F. Murtagh and A. Heck, (eds.), European Southern Observatory, Garching bei München, 423–428.

24. Pearl, J. (1984), *Heuristics — Intelligent Search Strategies for Computer Problem Solving*, Addison-Wesley, Reading, MA.

25. Rash, J. and Hughes, P., eds. (1988), *Proceedings of the 1988 Goddard Conference on Space Applications of Artificial Intelligence*, NASA Conference Publication 3009, NASA, Greenbelt, MD.

26. Rich, E. (1983), *Artificial Intelligence*, McGraw-Hill, New York.

27. Richmond, A., McGlynn, T., Ochsenbein, F., Romelfanger, F. and Russo, G. (1988), "The design of a large astronomical database system", *Astronomy From Large Databases: Scientific Objectives and Methodological Approaches*, F. Murtagh and A. Heck (eds.), European Southern Observatory, Garching bei München, 465–472.

28. Rosenthal, D. (1988), "Applying artificial intelligence to astronomical databases: a survey of applicable technology", *Astronomy From Large Databases: Scientific Objectives and Methodological Approaches*, F. Murtagh and A. Heck (eds.), European Southern Observatory, Garching bei München, 245–259.

29. Rosenthal, D., Monger, P., Miller, G., and Johnston, M. (1986), "An expert system for ground support of the Hubble Space Telescope", *Proceedings of the 1986 Goddard Conference on Space Applications of Artificial Intelligence and Robotics*, NASA, Greenbelt, MD.

30. Schank, R. (1987), "What is AI, anyway?", *AI Magazine*, 8, 59–65.
31. Sen, Amit (1988), private communication.
32. Smith, H. (1986), "Present and planned large groundbased telescopes: an overview of some computer and data analysis applications associated with their use", *Data Analysis in Astronomy II*, V. Di Gesù, L. Scarsi, P. Crane, J. Friedman and S. Levialdi (eds.), Plenum Press, New York, 141–154.
33. Steele, G. (1984), *Common Lisp*, Digital Press, Bedford, MA.
34. Tatar, D. (1987), *A Programmer's Guide to Common Lisp*, Digital Press, Bedford, MA.
35. Thonnat, M. (1988), "Toward an automatic classification of galaxies", *Le Monde des Galaxies, Physics*, H. Corwin and L. Botinelli (eds.), Springer-Verlag, New York.
36. Thonnat, M. and Clément, V. (1989), "OCAPI: An artificial intelligence tool for the automatic selection and control of image processing procedures", *Data Analysis in Astronomy III*, V. Di Gesù, L. Scarsi, P. Crane, J.H. Friedman, S. Levialdi and M.C. Maccarone (eds.), Plenum Press, New York, in press.
37. Tichy, W. (1987), "What can software engineers learn from artificial intelligence?", *Computer*, **20**, 43–54.
38. Turing, A. (1963), "Computing machinery and intelligence", *Computers and Thought*, E. Feigenbaum and J. Feldman (eds.), McGraw-Hill, New York, 1–35.
39. Vick, S. and Lindenmayer, K. (1988), "Verification and validation of rule-based systems for Hubble Space Telescope ground support", in *Proceedings of the 1988 Goddard Conference on Space Applications of Artificial Intelligence*, J. Rash and P. Hughes (eds.), NASA Conference Publication 3009, NASA, Greenbelt MD, 435–448.
40. Walker, J. (1988), "Supporting document development with Concordia", *Computer*, **21**, 48–59.
41. Winston, P. and Horn, B. (1988), *Lisp*, Addison-Wesley, Reading, MA.

3 Knowledge Based Telescope Scheduling

Mark D. Johnston
Space Telescope Science Institute
Homewood Campus
3700 San Martin Drive
Baltimore, MD 21218
USA

3.1 Introduction

The purpose of automating observatory scheduling is to increase the effective utilization and, ultimately, scientific return from one or more telescopes. The development of increasingly sophisticated satellite observatories, as well as the planned high level of automation of ground based telescopes, has led to a demand for more flexible scheduling so that astronomers can optimally exploit the capabilities offered by these facilities. Observing time on large telescopes has long been a very scarce resource: oversubscription by factors of several are typical, and this is likely to increase further in the future. It is thus important to consider how the utilization of existing and planned telescopes can be increased to the maximum extent possible.

Knowledge based scheduling is an approach to automated scheduling that exploits recent developments in "artificial intelligence" (AI) software technology. The aim is to build software systems that represent and reason with scheduling knowledge in a manner which imitates, at some level, the process carried out by human schedulers. This differs in approach from "classical", or operations research, approaches to scheduling, where the scheduling problem is cast into a mathematical framework for which exact or approximate methods of solution exist.

In the following we first consider the role of scheduling in the overall context of observatory operations in order to define the general characteristics of the observation scheduling problem. This is followed by a discussion of scheduling knowledge: how to represent and manipulate knowledge of scheduling constraints and preferences, and how to use this knowledge to guide the search for optimal schedules.

3.2 The Role of Scheduling in Observatory Operations

3.2.1 Modes of Observatory Operation

Several aspects of telescope operations are intimately related to scheduling, ranging from the initial allocation of observing time to the selection and sequencing of each night's exposures by the observer at the telescope. The extent to which automated scheduling can significantly increase observatory efficiency depends in large part on the "mode" in which the observatory is operated.

The simplest mode of telescope operation is the "classical" one: fixed blocks of time are pre-allocated to each program, with detailed scheduling of the block left up to the astronomer who travels to the telescope to make the observations. This mode has several advantages: advance planning for an observing run is straightforward, observers are in "control" of their own observations and, except for weather or other unpredictable factors, are primarily responsible for their quality. The drawbacks of this mode are well known: what can be observed is limited by the timing of the allocated block, and poor weather can cause an entire observing run to be lost. This mode has also tended to discourage long term programs which require a large number of (possibly non-contiguous) observations to accomplish their objectives.

The most obvious way to overcome these drawbacks is to move towards an "integrated" mode of operation in which time is not pre-allocated to specific programs but can be scheduled dynamically as conditions warrant ("flexible scheduling"). This mode is often referred to as "service" or "absentee" observing since observations are carried out by the telescope operations staff based on specifications provided by the proposing astronomer (see, e.g., Longair et al., 1986). Operating in an integrated mode permits observations to be optimally matched to the prevailing environmental and instrumental conditions. Not only does this promise to increase the effective utilization of the telescope, but it also increases the chances that any individual program will be carried out under its most favorable conditions. This latter point is particularly important for programs with the most restrictive observing requirements.

Some programs will continue to require the presence of the observer at the telescope, so that in practice some mixture of classical and integrated operation is likely to evolve. Remote observing (carried out the by the astronomer, but via remote control from a site more convenient than the telescope itself: Raffi, 1988) offers a promising way to combine these two modes to some degree, but only if remote observing stations become sufficiently widely available so that travel and advance planning for access to them can be minimized.

It is evident that the greatest efficiency gains from automated scheduling will be realized when the greatest flexibility exists to exploit good scheduling opportunities as they arise. In addition to the physical capability to respond

to schedule changes (e.g. rapid instrument changeovers) there are two main requirements on this flexibility:

- a sufficiently rich pool of candidate observations that a probable good match
  can be found to current environmental conditions;
- the absence of prior commitments that would forbid taking advantage of new
  and better scheduling opportunities when they occur.

These criteria are most easily met in an integrated observing mode where a
relatively large pool of observations is available to be scheduled at the discretion
of the telescope operations staff. Classical mode observing offers the fewest
opportunities for schedule optimization because of the generally limited number
of choices available to respond to varying observing conditions.

3.2.2 Scheduling and Operations

From an operations point of view, telescope scheduling can be roughly divided
into long term and short term problems. The long term problem is concerned
with construction of an overall scientific program for one or more observing
periods. This is essentially an "offline" process: it usually occurs before any
observations are executed, and requires only a relatively high level description
of the candidate programs, their resource requirements, and their most significant constraints (i.e. those with timescales of weeks to months). Short term
scheduling can be viewed as the process of deciding on a sequence of individual observations to schedule over a more limited time range. In general a short
term schedule represents an expansion in detail of some portion of a long term
schedule.

Long term and short term scheduling are of course intimately related. Both
are subject to the same basic telescope and operational constraints. Both schedule the same activities (but usually at different levels). It is important to devise
long term schedules which can be implemented when considered in detail, just as
it is important that short term schedules satisfy the overall boundary conditions
imposed by a long term schedule. These differences in timescale and level of
granularity are not fundamental: long term and short term scheduling can be
regarded as different perspectives on essentially the same process and can (and
should) make use of the same underlying scheduling knowledge.

Both long term and short term scheduling require a "knowledge base" of the
activities to schedule, i.e. their properties, relationships, priorities, etc. Since it
is undesirable for this to include explicit and redundant specifications for each
activity, some means must be provided for deriving implied properties from general activity descriptions. This amounts to a knowledge base describing the
observatory scheduling "context": for example, all observations made with a
particular instrument should automatically be associated with the general constraints on the use of that instrument, as well as with those on the telescope, the
visibility of the object to be observed, etc. Since this knowledge base is unlikely
to be static, some means must be provided for keeping it current.

This highlights the fact that effective observatory scheduling cannot be an
isolated process: for it to be ultimately useful it must be integrated into the

overall observatory operations environment. At the simplest level this integration must include the ability noted above to inform the scheduler of what must be scheduled. As implementation of the schedule proceeds, it is becomes equally important to provide up-to-date information about what has been executed and whether successfully or not. An even greater degree of integration should include capabilities for (Fosbury et al., 1988):

- automatic access to data on environmental conditions and predictions that can be used to update scheduling constraints;
- integrated planning capabilities, so that e.g. observations of various types are included in the scheduling pool along with appropriate calibration observations;
- user support for proposal preparation, including simulators and other observation design aids.

3.2.3 Characterizing Observatory Scheduling

Oversubscription: A major goal of flexible scheduling is to keep a telescope busy as much of the time as possible by executing the most scientifically productive observations. Since most observatories accept proposals for some specified time period (usually six months), flexible scheduling requires the acceptance of more observations for each period than can actually be accomplished, in order to account for the statistical variability of observing conditions during the period. Otherwise there is a risk of running out of appropriate observations, e.g. if there are significantly more clear nights than originally anticipated. The availability of a sufficiently rich pool of candidate observations is one of the major sources of efficiency gain to be expected from an integrated scheduling approach: the chances are higher that whatever conditions occur can be fully exploited. There are some implications of oversubscribing, however: some means (e.g. priority) must be established for selecting among equally appropriate observations, and for indicating whether it is more important to complete a particular program in progress or to start a new one.

Constraint variety: Observatory scheduling is characterized by a large number of constraints arising from many different sources, where a constraint is any factor that affects when an activity can or should be scheduled. Some constraints are related to the objects to be observed and their visibility with respect to the horizon, Sun, and Moon. Some are derived from environmental conditions, e.g. constraints on seeing or sky background light. Others come from operational characteristics of the telescope and instrumentation, e.g. constraints related to slewing, instrument reconfiguration, and calibration. Others come from required relationships among the exposures to be taken, e.g. time precedence and separation requirements. Still others come from human factors, e.g. not overloading the observatory staff, or disrupting the travel plans of an observer travelling to the telescope. It is important to note that, because of constraint interactions, these various factors must be considered effectively simultaneously.

Extreme range of constraint timescales and effects: Not only do constraints come from many sources, but they operate and vary their impact on many different timescales. The most extreme manifestation of this is probably to be found in low Earth orbit satellite observatories, where the dominant constraint timescales can vary from a small fraction of the orbital period up to a year or more. Constraints can also vary drastically in effect, from indicating a slight preference for one time over another to abruptly prohibiting the scheduling of some activities under certain conditions.

Varied optimization criteria: Since the primary purpose of automating telescope scheduling is to "optimize" telescope utilization, it is clearly important that schedule optimality be defined. This is less straightforward than it might seem at first: scheduling goals vary depending on the circumstances, so that a schedule which is optimal in some sense can be far from optimal in another. For example, at different times the most important optimization criterion could be one of maximizing overall telescope throughput, picking up a disrupted schedule, diagnosing an instrument problem, or scheduling a best match to changing environmental conditions.

Unpredictability: This is by no means peculiar to astronomical observation scheduling but is a general characteristic of scheduling real-world activities. While some (e.g. celestial) constraints are predictable with extremely high accuracy, other types of constraints vary widely in their degree of predictability. On the ground, weather is the single most significant unpredictable scheduling factor, although other factors (e.g. breakdowns) can obviously be important. Satellite observatories do not suffer from the vagaries of the weather, but usually have their own unpredictable constraints: e.g. whether necessary communications pathways will be available, or whether selected guide stars can be acquired. Unpredictable constraints limit the scheduling "horizon", i.e. how far into the future accurate schedules can be constructed.

3.3 Approaches to Automated Scheduling

Computer techniques for optimal scheduling have been investigated for many years for a number of applications (see, e.g., King and Spachis, 1980, for a comprehensive review and bibliography). Much of this classical work has focused on versions of the idealized "job shop" scheduling problem, i.e. the problem of scheduling n tasks on m machines. This problem and related ones are NP-complete, meaning essentially that there are no efficient algorithms for finding optimal solutions (see, e.g., Garey and Johnson, 1979).

The basic problem with classical approaches is that they require key features of the problem to be abstracted away, so that even an "exact" solution to the abstracted problem is often of limited relevance to the original "real" problem. Approximate solutions to the abstracted problem suffer from the same limitations. For example, there exist good approximate methods for finding near-optimal solutions to the well-known "travelling salesman" problem, to which telescope scheduling reduces if minimizing target-to-target slew time is the *sole* scheduling criterion. But this is rarely the case: other constraints enter into the problem in an essential way, and these generally cannot be formulated within the framework of the abstracted problem.

It is clear that classical approaches can be useful for problems which are sufficiently simple: in practice this often means that schedule optimization is driven by a *single* overriding criterion. For the problem of scheduling complex modern observatories, however, this will rarely be the case: more powerful techniques are required.

In recent years a variety of new software methodologies have been developed under the general term of "artificial intelligence" (AI). This refers to a collection of software development techniques and tools that have evolved in the course of computer science research as effective ways to represent and solve certain kinds of problems. These techniques have moved from the laboratory into widespread use in applications as their effectiveness has been demonstrated. For the purposes of automated scheduling, the most important of these are: a language (Lisp) that is particularly appropriate for manipulating complex data structures and symbolic data; object oriented programming with inheritance; rule based programming facilities; and new techniques for heuristic search.

Several artificial intelligence research efforts have considered scheduling as a domain where AI techniques can be fruitfully applied. Of particular interest is the factory scheduling work of Fox, Smith, and co-workers (Fox, 1983; Fox and Smith, 1984; Smith et al., 1986) who have developed a rich constraint representation and search methodology for attacking realistic factory scheduling problems. While factory scheduling shares a number of common features with telescope scheduling (most notably a similar set of precedence and efficiency constraints), there are some major differences. Certain important factory scheduling constraints (e.g. minimizing inventory) are not relevant for telescope scheduling, while telescope scheduling is characterized by a large number of constraints with extreme ranges in timescales.

At the Space Telescope Science Institute (STScI) we have for some time been working on a project (SPIKE) to apply AI software technology to the problem of scheduling the NASA/ESA Hubble Space Telescope (HST; see Hall, 1982; and ref. 2). Space Telescope scheduling is an extremely demanding task (Miller et al., 1987; Johnston, 1988a; Miller et al., 1988; Johnston and Miller, 1988), requiring the scheduling of some tens of thousands of exposures per year. These exposures are subject to a large number of scheduling constraints, some derived from the scientific goals of the proposer, some a consequence of HST design, operating characteristics, and low Earth orbit environment. The most extended HST schedules will cover a year or more and will ensure that the basic plan for implementing the HST science program is feasible and balanced. Shorter term schedules are used to lay out detailed observation sequences and make requests for communications contacts. The most detailed schedules specify spacecraft and instrument commands and the times they are to be executed. Because HST operates almost entirely in a pre-planned mode, detailed short term schedules must be fully specified weeks ahead of time. These schedules are integrated, in that exposures from many different proposals may be scheduled to occur during a single day of observing.

The initial phases of SPIKE development have concentrated on a general framework for representing and reasoning with scheduling knowledge, and the application of this framework to the long term HST scheduling problem. The framework is described in the following sections.

3.3.1 Reasoning with Scheduling Knowledge

3.3.1.1 Types of Scheduling Knowledge. Before considering how a computer could represent and reason with scheduling knowledge it is useful to consider what form this type of knowledge generally takes when expressed in everyday terms. How would a human scheduler evaluate possible choices when given a partially completed schedule and asked what scheduling decision should be made next? More specifically, how would a human scheduler evaluate the scheduling possibilities for a single activity?

Suppose we are concerned with scheduling an activity A_i over some time period, given that activities $A_j, j \neq i$, are already scheduled at times t_j. A human scheduler would assess the opportunities for scheduling A_i at various times by considering the implications of the known constraints on A_i. For example:

- don't schedule A_i on any Sunday
- it's better to schedule A_i during the first week of a month than the second
- don't schedule A_i to follow A_j
- A_i cannot be scheduled to be simultaneous with any of $A_j, A_k, \ldots$
- schedule A_i to happen as soon as possible after A_k, but no more than a week later
- schedule A_i at times which avoid contention for a scarce resource used also by $A_m, A_n, \ldots$

We can generalize statements of this type to a more abstract form as follows:

Given that $A_j, \ldots$ are scheduled at $t_j, \ldots$, the "degree of preference" for scheduling A_i at t due to constraint p is $W_i^p(t; t_j, \ldots)$

where the dependence on $t_j, \ldots$ can be dropped for constraints that depend only on time, i.e. that are independent of when other activities are scheduled. "Degree of preference" can be assigned in some numerical range corresponding to the value judgment involved, with larger values for W corresponding to greater preference. W can represent either a deterministic constraint or some classes of intrinsically unpredictable constraints, e.g. if W can be formulated in terms of the probability that desirable conditions will be met.

3.3.1.2 Combining Scheduling Constraints and Preferences. To assess the possible scheduling times for A_i it is necessary to combine in some manner the "degrees of preference" derived from all applicable constraints. This combination process is formally similar to that employed in rule based expert systems which assess evidence for and against various diagnostic hypotheses (e.g. MYCIN, Shortliffe, 1976; and Prospector, Hart et al., 1978). For example, the MYCIN expert system for medical diagnosis used rules with associated "certainty factors" such as:

```
IF (1) the stain of the organism is gramneg, and
   (2) the morphology of the organism is rod, and
   (3) the patient is a compromised host
THEN there is suggestive evidence (0.6) that the identity
   of the organism is pseudomonas
```

Knowledge of each rule antecedent contributes a certain "weight of evidence" for the rule conclusion. The certainty factor of the conclusion is derived from the antecedents by a combination function $F(CF_1, CF_2, \ldots)$. A variety of combination functions have been proposed, but it has been shown that, under plausible conditions on how the combination functions behave, all of these methods are equivalent in that they can be transformed to (and from) the addition of real numbers (see, e.g., Hajek, 1985; Cheng and Kashyap, 1988).

All of the scheduling "evidence" provided by the constraints could be combined with one of the functions used in other rule based expert systems, but this would ignore an important aspect of scheduling constraints: they often specify times when an activity *cannot* be scheduled without violating a strict constraint. In particular, it is desirable to ensure that if there is overwhelming evidence from *any* constraint against scheduling an activity at a certain time, then no amount of evidence in favor of that time from other constraints can override this fact. This effect can be obtained by taking the exponential of the additive degree of preference W and adopting a multiplicative form of evidence combination. In the limit as the additive degree of preference W goes to minus infinity (i.e. overwhelming evidence against scheduling at some time), the exponential goes to zero, and so the product with any other exponential preference $B = \exp W$ is also zero. When W is finite, B is positive; just as for W, larger values of B indicate greater degrees of preference.

3.3.1.3 Suitability Functions. Before this approach can be used in a practical setting, some means must be found to handle the multidimensional nature of the $B_i^?(t; t_j, \ldots)$ functions. The problem can be simply illustrated by a binary constraint representing a precedence relationship between two activities, e.g. that A_j must follow A_i. The two-dimensional B function for this constraint on A_j can be written as:

$$B_j(t; t_i) = \begin{cases} 1 & \text{if } t > t_i + d_i; \\ 0 & \text{otherwise} \end{cases}$$

where d_i is the duration of activity i. Until A_i is fixed at a specific time t_i it is generally not possible to say with certainty whether a candidate time t for A_j is allowed or forbidden. We can, however, "project" the two-dimensional function B to a one-dimensional function as follows:

$$S_j(t) = \max_{t_i} B_j(t; t_i)$$

where the max is taken over all times when A_i is permitted to be scheduled. S is called a "suitability function" and has the following desirable properties:

- $S_j(t) = 0$ if and only if *no* possible choices for scheduling A_i will allow A_j to be scheduled at t,

- otherwise its value is the *highest* possible value of B that can be obtained, regardless of when A_i is scheduled.

The former property is important since from S we can identify times that are certainly excluded for scheduling A_j. The latter property is also important, since to the extent that schedule optimality is indicated by higher values of B, higher values of S indicate better choices for scheduling A_j, i.e. the resulting overall schedule will be better. Although formulated here for binary constraints (those involving only two activities), the generalization to higher-order constraints is straightforward (Johnston, 1989).

Each constraint on an activity determines a suitability function as defined above. From the multiplicative combination of B, the total suitability function for an activity A_i is the product of the suitabilities from all of its constraints:

$$S_i(t) = \prod_p S_i^p(t)$$

Since $S_i(t) = 0$ only if A_i cannot be scheduled at t (due to one or more constraints), the allowed scheduling times for A_i are limited to those times when $S_i(t) \neq 0$ and to times which have not been excluded by any scheduling decision. Thus the definition of suitability in terms of $\max B$ over permitted times becomes a set of implicit equations for the suitability functions of each activity to be scheduled. These equations can be solved iteratively after each scheduling decision.

The suitability function framework described here was developed for the HST SPIKE scheduling software to represent constraints and preferences on the scheduling of Space Telescope observations. However the framework itself is completely general: once a constraint is formulated in terms of "degrees of preference" then there is a straightforward prescription for turning it into a suitability function for consideration along with all other relevant constraints. The advantages provided by the suitability function framework are several:

- a simple and uniform way to capture human value judgments, preferences, and trade-offs among conflicting constraints;
- a mechanism for propagating constraints, i.e. deducing the joint consequences of constraints and strategic scheduling decisions;
- a means for avoiding some scheduling decisions which would later have to be retracted;
- a way to compare different times for scheduling activities in light of all of the preferences that bear on the possible choices;
- extensibility, in that the set of constraints can be easily extended or modified by adding new constraints or re-defining the effects of existing ones;
- a mechanism for tracking the probable and/or certain consumption or loading of critical resources;
- explanatory capabilities: it is easy to see the effects of all constraints on an activity by inspecting their contributions to the total activity suitability.

For computational reasons it is necessary to discretize the essentially continuous formulation above: in SPIKE this was done by restricting suitability functions to be piecewise constant functions of time. This has the advantage of closure under the most common operations (i.e. the product of two piecewise constant functions is piecewise constant) and time/storage efficiency (proportional to the number of time intervals with distinct values). Alternatively, it would be possible to discretize the time domain if there were an appropriate natural timescale for the problem.

3.3.2 Using Scheduling Knowledge

3.3.2.1 Scheduling as Search. The process of constructing a schedule is essentially a search problem: at each step, the scheduler must choose an activity, then choose how to restrict the times when it can be scheduled. Since there are usually a large number of possible choices for the former, and a generally infinite number of choices for the latter, the effort required to exhaustively search the space of possible schedules is typically exponential in the size of the problem. This is impractical for all but the smallest cases and exemplifies the characteristic "combinatorial explosion" of NP-complete problems.

An important aspect of the search problem for astronomical observation scheduling is that there is often no solution because the schedule is intentionally oversubscribed. It is generally infeasible to enumerate all possible deadends in order to prove that this is the case. It is therefore necessary to provide the scheduler with knowledge about appropriate stopping criteria, i.e. when a partial schedule can be considered satisfactorily complete even though not all activities have been scheduled. For example, some programs may have to be scheduled in their entirety in order to meet their scientific objectives: scheduling only part of the program is therefore not acceptable. Others could have "completion thresholds", e.g. that a specified percentage of the observations in the program must be scheduled. Still others could be lower priority and scheduled only to fill gaps around high priority observations.

In addition to oversubscription there is another source of activities which can turn out to be inherently unschedulable: these are related by mutually inconsistent sets of constraints. When designing complex observing programs it is not difficult to inadvertently overconstrain some observations in the program by specifying constraints that cannot be simultaneously satisfied by any schedule. As a result, it is important to exploit techniques that can identify and diagnose sets of overconstrained activities without becoming bogged down in fruitless search. Diagnosis involves the identification of mutually inconsistent sets of constraints, some of which could be relaxed or removed (but usually only by redesigning some portion of the observing program). The fact that this problem has been encountered in a number of HST observing programs suggests that proposers could benefit from observation design aids that include a scheduling component (see also Fosbury et al., 1988).

A central concept in knowledge based scheduling is that of using constraints to direct the search process. In the suitability function framework, constraint knowledge is embodied in the suitability functions $S_i(t)$ of the activities A_i to be scheduled. This knowledge can be exploited to limit search in two major ways:

1. all times when $S_i(t) = 0$ can be immediately excluded as in violation of a strict constraint;

2. the value of $S_i(t) > 0$ can be used to select among alternative choices for times to schedule activities: higher values of $S_i(t)$ indicate greater satisfaction of combined preference constraints.

Suitability functions and the constraints which contribute to them can also be used to construct heuristics for which activities to select next for scheduling, e.g.:

- activities with the least total time where $S_i(t) > 0$ (or, more generally, where $S_i(t)$ is greater than some threshold);
- activities related by the most restrictive constraints (i.e. the smallest intervals where $B_i(t) \neq 0$);
- activities participating in the largest number of constraints;
- activities requiring the most heavily subscribed resources.

Note that it is not necessary to fix a selected activity at a specific time: a scheduling decision could be to restrict the scheduling times of an activity to a subset of those allowed by the constraints. E.g. times when $S_i(t)$ is less than some threshold could be eliminated, thus retaining only the "best" possibilities for future consideration.

A great deal of research has been conducted into general strategies for minimizing the amount of backtracking required in search (see, e.g., Mackworth, 1977; Dechter and Pearl, 1988). Most of these strategies were developed for discrete constraint satisfaction problems, in contrast to scheduling where continuous variables (the times to schedule activities) are involved. However it is possible to apply some of these techniques to the continuous domain, and some are implicit in the suitability function framework. A detailed review of these methods is beyond the scope of this discussion; see Johnston (1989) for details.

3.3.2.2 Backtracking and Schedule Repair. Effective search requires the early identification of both "good" decision paths as well as the early pruning of "deadend" paths, i.e. partial schedules must be judged by their *potential* for being completed beneficially as well as by their current state. This is complicated by the fact that scheduling conflicts may not be detected until many steps into the search, at which point a large amount of effort may have already been expended. In this case, backtracking, i.e. "undoing" some previous decision, is required in order to extend the schedule. In its simplest form, backtracking would simply undo some number of the most recent decisions, then make a scheduling choice different from one originally made. This has the drawback that the intervening decisions may have nothing to do with the conflict, and thus the effort expended on them is lost. It is therefore desirable if possible to identify a minimal number of past decisions to undo to resolve the conflict and thus repair the partial schedule.

Again, suitability functions and the constraints contributing to them can be used to select decisions to undo. When a deadend is encountered, i.e. some activity cannot be scheduled at any time, then from the constraints contributing to that activity's suitability it is possible to identify which other activities could be responsible for the deadend. The scheduler can then undo decisions only for the related activities, leaving untouched any intervening choices which can have no effect on the problem.

3.3.2.3 Hierarchical Scheduling. A common and important problem solving strategy is to formulate and solve a simpler higher level problem, then attack the resulting lower level subproblems by constraining them with the higher level solution. In the scheduling domain there are two obvious ways to accomplish this: by scheduling *groups* of related activities at once, and by limiting the *time granularity* of the schedule.

For example, it is possible to cluster appropriately related activities into a single "meta-activity" which can be scheduled initially as a single entity. Activities with similar constraints could be considered candidates for this type of clustering. A typical example of this in observation scheduling would be to schedule observations of the same or nearby targets with the same instrument to occur together, unless there are constraints which would force them to be separated. In this case single decisions can refer to an entire group of activities at once.

Hierarchical scheduling in time involves dividing the overall scheduling interval into subintervals, so that activities (or clusters of activities) need initially be committed only to one of the subintervals. How the interval is divided depends on the details of the problem, but care should be taken that the division does not artificially eliminate any scheduling opportunities that could ultimately prove important. To take an example from HST scheduling, long term schedules of a year or longer duration can be practically divided into subintervals of duration $\sim$ 30 days, then subdivided further into $\sim$ 10 day intervals. Activities are initially committed to the longer subintervals, then to the shorter.

This approach has the major advantage that some constraints which are important at the detailed scheduling level can be treated in an average or statistical sense when activities are allocated only at a sufficiently coarse-grained level. This will generally further simplify the calculation and propagation of constraints. In this case the limit on the schedule granularity follows from the limits on the validity of the statistical constraint approximation. For example, Space Telescope has a large number of constraints that relate to target visibility during each orbit: as long as intervals much greater than the orbital period are considered, these constraints can be summarized as average visibilities during the intervals. This allows the detailed orbital constraints to be ignored during long term scheduling, with the assurance that, in a statistical sense, the required target visibility constraints will be satisfiable at a detailed level.

The drawback of hierarchical scheduling is that levels may destructively interact: a deadend in a detailed schedule may require revising the higher level schedule, which can potentially invalidate other detailed schedules. It is thus important that the higher level problem reflect as accurately as possible the constraints that will be important in detailed scheduling.

3.3.2.4 Scheduling and Unpredictability. Dealing with unpredictability is one of the most difficult aspects of scheduling in complex environments. There are two basic approaches for dealing with unpredictability, characterized by whether the scheduling effort is focused on *anticipating* or *reacting* to unpredictable occurrences. Which approach is most appropriate for any given scheduling problem depends on the timescales and impacts of the unpredictable constraints, and on their typical effect on the remainder of the schedule.

Typical anticipatory strategies include the following:

- Minimize the probability of schedule failure: If an unpredictable scheduling factor can be characterized statistically, then it is often possible to formulate a constraint which expresses the goal of scheduling activities at times which minimize the probability of schedule disruption. For example, most Hubble Space Telescope exposures require a pair of guide stars for pointing stability, but it cannot be known with certainty in advance whether a pair selected on the ground can actually be acquired. It is, however, possible to estimate the probability of successful acquisition, based on the expected frequency of situations when attempted acquisitions will fail. With this knowledge a constraint can be defined to favor the scheduling of exposures at times which minimize the probability of failure. This type of constraint can be easily expressed as a suitability function whose value is directly related to the probability that desirable events will occur.

- Optimize the schedule for robustness: It is often possible to minimize the effect of unpredictable events by appropriately ordering activities in the schedule. For example, a fruitful strategy for scheduling critical highly-constrained observations is to schedule them as early as permitted by their constraints, and then schedule lower priority observations at other times where the critical observation could be done. Then if the critical observations cannot be executed when originally scheduled, there remain opportunities for shifting it to another time, with minimal impact on the remainder of the schedule.

- Construct alternative schedules: When the range of variation of the unpredictable constraint is not too great, it is often possible to schedule for alternative possibilities in advance and then select the most appropriate alternative when necessary. For example, in ground based telescope scheduling, it may be possible to construct several alternative schedules based on short term weather predictions, then select the best match to the actual conditions when they are known. Such alternative schedules can be open-ended or can be constrained to re-join a "master" schedule at some specified time. An example of the latter might be scheduling of "backup" observations to be executed only if a primary observation cannot be done for some reason.

An alternative to anticipatory scheduling is reactive scheduling (see, e.g., Ow et al., 1988). At one extreme are pure dispatching strategies and heuristics, i.e. no predictive scheduling is carried out at all. These strategies are successful in extremely dynamic environments when the likelihood is small that any predictive schedule can be followed. Less extreme approaches involve modifying existing schedules in response to changes. Various strategies can be employed, including removing some already scheduled activities to make room for "bumped" activities, or shifting some activities to other times.

It can be expected that most observatory scheduling problems will benefit from some combination of anticipatory and reactive scheduling. There will always be a need for pure reactive scheduling when totally unexpected events occur (e.g. instrument breakdown), but more efficient operation is likely when a predictive schedule is developed and followed to the greatest extent possible.

A detailed predictive schedule may be invalidated by unpredictable constraints, but a schedule developed at a higher level of abstraction may remain valid even in the face of unpredictability at lower levels. For example, it is not possible to know *when* in each month the nights with best seeing will occur, but a schedule at the month level of time granularity may nevertheless remain valid if the expected *number* of nights are available. This argues for the maintenance of schedules at multiple levels, with higher level schedules acting continuously to constrain lower level ones.

3.3.3 The Human Interface to Scheduling Systems

An important aspect of any scheduling software system is how it interacts with the people who use it. Users must have visibility into all aspects of the scheduling problem and the evolving schedule. They must also ultimately have control, i.e. the ability to override any decisions made by the scheduling software, and the ability to create and evaluate alternative schedule fragments. In large part this is because even the most well-intentioned attempt to capture *all* of the knowledge required to construct realistic schedules will remain incomplete. Scheduling problems are dynamic: new constraints arise, old ones need to be modified, and exceptions to expectations inevitably occur.

Because of the large volume of information required to specify even modest-sized realistic scheduling problems, it is almost essential to utilize graphical display and interaction capabilities. This leads to the concept of implementing scheduling tools on single-user workstations, where high-speed graphics windows and dedicated processing power can both be exploited. This approach has been employed in the HST SPIKE scheduling software, where multiple screen windows can be constructed by the user to graphically display aspects of one or more schedules (e.g. the initial and current impact of all constraints on an activity). Using a mouse, the user can control the operation of the system, e.g. by invoking various automatic scheduling algorithms, or manually making (and unmaking) any scheduling decision. Even with this approach, considerable software is required to help the user determine what is interesting to display.

3.4 Discussion and Future Prospects

The essential applicability of knowledge based techniques to astronomical observation scheduling is demonstrated by development of the HST SPIKE scheduling software. AI development methodology, tools, and approaches are clearly appropriate for these kinds of complex scheduling problems. SPIKE is currently being used to schedule observations for HST ground system testing. It is also expected to play an essential role in the feasibility evaluation of HST proposals, i.e. ascertaining whether proposed observations are schedulable.

Since one of the major advantages of the knowledge based approach is extensibility, SPIKE has been applied to prototypical scheduling problems for two other observatories: the European Southern Observatory 3.6m ground based telescope in Chile (Johnston, 1988b), and the long term scheduling of the IUE

observatory (i.e. the allocation of observing shifts to weeks; Johnston, 1988c).
In both cases SPIKE was easily adapted to the new problems simply by adding
the appropriate activities types and constraints.

Future development of the knowledge based approach can be expected in
several directions:

- Scheduling knowledge acquisition: although a library of observing constraints
  has been created for SPIKE, the definition of a new constraint still requires
  a programmer to create the new constraint type and specify its behavior. It
  would be desirable for users to be able to do this themselves, either via a
  constraint "language" or through graphical constraint editing tools.

- Smarter search algorithms: there remains significant room for improvement
  in the design of algorithms to exploit constraint knowledge for more efficient
  search. Among the most promising approaches under investigation are "artificial neural networks" (Adorf and Johnston, 1989; Johnston and Adorf,
  1989); other approaches include rule based strategies (Miller et al., 1988)
  and genetic algorithms. It remains to be seen which approaches will prove
  most fruitful for the widest range of scheduling problem types.

- Full integration with observatory operations: as discussed above in section
  3.2.2, the full benefits of automated scheduling can only be realized when
  updated constraint knowledge and feedback from schedule execution is readily available to the scheduler. Recent advances in workstation architectures
  and networking capabilities are encouraging in this regard.

- Multi-observatory scheduling: the planning of coordinated programs among
  different observatories (e.g. multi-wavelength campaigns) can be very complex: the constraints for every participating observatory must be considered
  simultaneously. Knowledge based scheduling systems such as SPIKE can be
  readily adapted to this type of problem.

Acknowledgements. The author is grateful to G. Miller, J. Sponsler, S. Vick,
K. Lindenmayer, and R. Jackson (STScI), D. Rosenthal (NASA Ames Research
Center) and H.-M. Adorf (ST-ECF) for many useful discussions, and for the hos-
pitality and support of the Space Telescope — European Coordinating Facility
and the European Southern Observatory (Garching) where some of this work
was conducted. Space Telescope Science Institute is operated by the Association of Universities for Research in Astronomy for the National Aeronautics and
Space Administration.

References

1. Adorf, H.-M. and Johnston, M. (1989), "A discrete stochastic 'neural network'
   algorithm for constraint satisfaction problems", in preparation.
2. *Call for Proposals*, October 1985, Space Telescope Science Institute, Baltimore, MD.

3. Cheng, Y. and Kashyap, R. (1988), "An axiomatic approach for combining evidence from a variety of sources," *Journal of Intelligent and Robotic Systems*, **1**, 17–33.

4. Dechter, R. and Pearl, J. (1988), "Network-based heuristics for constraint satisfaction problems," *Artificial Intelligence*, **34**, 1–38.

5. Fosbury, R.A.E., Adorf, H.-M. and Johnston, M. (1988), "VLT operations — the astronomers' environment," in *ESO Conference on Very Large Telescopes and their Instrumentation*, Vol. 2, M.-H. Ulrich (ed.), European Southern Observatory, Garching bei München, 1283–1290.

6. Fox, M. (1983), "Constraint-directed search: A Case Study of Job Shop Scheduling," Ph. D. Dissertation, Computer Science Department, Carnegie-Mellon University. Revised version published 1988 by Morgan-Kaufmann, Los Altos, CA.

7. Fox, M. and Smith, S. (1984), "ISIS: A knowledge based system for factory scheduling," *Expert Systems*, **1**, 25–49.

8. Garey, M. and Johnson, D. (1979), *Computers and Intractability*, W.H. Freeman & Co., San Francisco.

9. Hájek, P. (1985), "Combining functions for certainty degrees in consulting systems," *International Journal of Man Machine Studies*, **22**, 59–76.

10. Hall, D.N.B. (1982), *The Space Telescope Observatory*, Special Session of Commission 44, IAU 18th General Assembly, Patras, Greece, NASA CP-2244.

11. Hart, P., Duda, R. and Einaudi, M. (1978), "Prospector — a computer-based consultation system for mineral exploration," *Mathematical Geology*, **10**, 589.

12. Johnston, M. (1988a), "Automated telescope scheduling," *Coordination of Observational Projects in Astronomy*, C. Jaschek and C. Sterken (eds.), Cambridge University Press, Cambridge UK, 219–227.

13. Johnston, M. (1988b), "Automated observation scheduling for the VLT", in *ESO Conference on Very Large Telescopes and their Instrumentation*, Vol. 2, M.-H. Ulrich (ed.), European Southern Observatory, Garching bei München, 1273–1282.

14. Johnston, M. (1988c), "Artificial intelligence approaches to spacecraft scheduling," in *Proceedings of the ESA Workshop on Artificial Intelligence Applications for Space Projects*, (c/o A. Scheffer, Mathematics and Software Division, ESTEC, Noordwijk), Nov. 1988, 5 pp.

15. Johnston, M. (1989), "Reasoning with scheduling constraints and preferences," in preparation.

16. Johnston, M. and Miller, G. (1988), "Artificial intelligence approaches to astronomical observation scheduling," *Data Analysis in Astronomy III*, V. Di Gesù, L. Scarsi, P. Crane, J.H. Friedman, S. Levialdi and M.C. Maccarone (eds.), Plenum Press, New York, in press.

17. Johnston, M. and Adorf, H.-M. (1989), "Scheduling with neural networks", in preparation.

18. King, J.R. and Spachis, A.S. (1980), "Scheduling: bibliography and review," *International Journal of Physical Distribution and Materials Management*, **10**, 105–132.

19. Longair, M.S., Stewart, J.M. and Williams, P.M. (1986), "The UK remote and service observing programmes," *Quarterly Journal of the Royal Astronomical Society*, **27**, 153.

20. Mackworth, A. (1977), "Consistency in networks of relations," *Artificial Intelligence*, **8**, 99.

21. Miller, G., Rosenthal, D., Cohen, W. and Johnston, M. (1987), "Expert system tools for Hubble Space Telescope observation scheduling," in *Proceedings of the 1987 Goddard Conference on Space Applications of Artificial Intelligence*; NASA, Goddard Space Flight Center, Greenbelt, MD; reprinted in *Telematics and Informatics*, **4**, 301 (1987).

22. Miller, G., Johnston, M., Vick, S., Sponsler, J. and Lindenmayer, K. (1988), "Knowledge based tools for Hubble Space Telescope planning and scheduling: constraints and strategies", in *Proceedings of the 1988 Goddard Conference on Space Applications of Artificial Intelligence*, NASA CP 3009, Goddard Space Flight Center, Greenbelt, MD, 91–106.

23. Ow, P, Smith, S. and Thiriez, A. (1988), "Reactive plan revision," in *Proceedings of the American Association for Artificial Intelligence 1988*, 77.

24. Raffi, G. 1988: "Remote observing," in *ESO Conference on Very Large Telescopes and their Instrumentation*, Vol. 2, M.-H. Ulrich (ed.), European Southern Observatory, Garching bei München, 1061–1072.

25. Shortliffe, E. (1976), *Computer-Based Medical Consultations: MYCIN*, American Elsevier, New York.

26. Smith, S., Fox, M. and Ow, P. (1986), "Constructing and maintaining detailed construction plans," *AI Magazine*, Fall 1986, 45

The Data Analysis Process

4 Survey Work with Automated Data Analysis

D. Teuber, P. Schuecker and H. Horstmann
Astronomisches Institut
Westfälische Wilhelms-Universität
Wilhelm-Klemm-Str. 10
D-4400 Münster
F.R. Germany

4.1 Introduction

During recent years the detection of large scale structures in the Universe has become technically feasable. High quality photographic sky surveys, fast and precise measuring machines and sufficient computing power were the prerequisits. The Muenster Redshift Project (MRSP) described by Horstmann (1988) and Schuecker (1988) investigates the distribution of luminous matter up to $z = 0.3$. The large amounts of data to be processed and interpreted require the support of an Astronomical Data Analysis System (ADAS) suited for that purpose. The hardware of the ADAS has been briefly introduced by Teuber (1988b). Here, some aspects of the software environment and two applications will be discussed.

Knowledge about the distribution of different types of objects is basic to topological studies of complex astronomical structures. In the MRSP the relevant objects are galaxies, constituting clusters of galaxies, superclusters and possibly higher-order structures. The distributions of different types of stars contain structural information about our Galaxy. In order to detect sufficiently many objects within reasonable time, wide-angle (36 square degrees) photographic plates are digitised with the specially equipped microdensitometer PDS 2020GMplus. At high galactic latitude about 150,000 objects (stars and galaxies) are found on a direct plate and about 50,000 spectra on an objective prism plate. The study of several hundred plate pairs (direct and prism) with a total number of several million objects is necessary for detecting the complex astronomical structures mentioned above. The survey project has been divided into four distinct components which are listed in Table 4.1.

Table 4.1. Components of the survey project and their performance specifications.

MRSP Survey Components
Process and database management
– user/computer interaction
– easy adaptation of computer programs to new problems
– fast data handling
– survey protocol
Data acquisition
– reliable data acquisition machines
– minimum of machine-dependent biases
– steady data flow
Object isolation/identification
– minimum of machine-dependent biases
Object reduction/classification
– minimum of machine-dependent biases

Large survey work relies on the economical measurement and reduction of a very large number of photographic plates. The astronomical parameters of the individual objects have to be extracted from the complex pattern of the photographic plate. This is an abstraction process, which comprises segmentation, preprocessing and classification of the individual astronomical objects. A process is called "automated abstraction" when the level of abstraction is determined by a piece of software. Only automated abstraction can supply the large amount of data. Automated abstraction relies on automated data acquisition, reduction and database administration.

Application software for automated abstraction must be implemented such that the sequence of algorithms can be modified easily and quickly. In many cases no standard procedures for the actual survey recognition task exist. The optimal procedures can often only be found by trial and error.

The storage format of control parameters and data for automated abstraction must be chosen to be very efficient in relation to the computer in order to maintain speed. This constraint contradicts a user friendly format. So a tool has to be introduced, to make control information for human users accessible and to make visible the mutual interactions of the individual components of the ADAS.

Software maintenance problems arise because survey work is carried out on a long term basis. A maximum of transparency regarding the functions of the individual algorithms applied is needed. Application software must be portable to allow the change of computer systems.

4.2 Generic Applications and Monitors Enviroment

GAME has been developed and implemented at Muenster to provide the infrastructure for interactive data reduction and the MRSP expert system. A first concept of GAME was introduced by Teuber (1985). Reports by Teuber (1988a; 1988b) are concerned with the internal structures of GAME. Also, considerations are included regarding the requirements of an ADAS in an area where applications are used in classical (interactive and batch) and automated modes. GAME has been influenced by the ideas of abstract data types and generic definitions (Shaw, 1984). The recent development of GAME is directed towards object oriented programming (Goldberg and Robson, 1983).

In object oriented programming an object is defined as a set of information together with the description of how the information can be manipulated. The operations are called methods. Objects communicate by sending each other messages. An object executes a method when it receives an appropriate message. A message specifies WHAT operation shall be performed. HOW the request is fullfilled is part of the definition of the object.

GAME supplies tools for handling information in an object oriented way and offers support for program development. GAME is the link between all parts of the ADAS, hardware and software. GAME screens the users and applications from the specific properties of the actual computer system, as is usual for all modern data analysis facilities. GAME helps the user to look at his data and procedures on a more abstract level than the file system level. The integration of Lisp application programs is currently under study.

The computer system currently employed is bound to Fortran. Therefore, the object oriented approach concerns only the structures that are definable by the programmer. Still, this approach offers advantages. The interfaces to GAME (GIF) are a few messenger routines. Particular demands of users can be satisfied by adding new classes of objects into GAME without touching the rest of the system.

The object classes that may be used by application software are built from a set of internal classes. We will introduce the applicative object classes which are used for scientific data. Details on the internally used classes can be found in the GAME System Manual (Teuber, 1989).

Data are organised by GAME in tree structures forming the hierarchical database (HDB). The branches are constructed from directory nodes which are instances of the respective class. The leaves are objects of other classes. On this basis the applicative objects are built. On the application level four classes of objects are available. Using the appropriate interface calls, the application programs (AP) may store, retrieve and manipulate objects of classes **parameter**, **table**, **image** or **SIP** (single pattern).

Objects of class **parameter** are the simplest units to process data. A parameter is identified by the name of its node. It addresses a storage area that is divided into a control and a data segment. The former holds the description of the data which are stored in the latter. A structural description allows the data value(s) to be accessed as a single value, one dimensional array, stack, ring buffer or menu.

Objects of class **table** can be used to process tables and catalogues of various formats using a corresponding definition file. Once the definition file is created, all GAME users are able to access the respective catalogue. A native GAME table format has not yet been specified. Instead, user defined formats are supported.

Objects of class **image** are used to process the pixel map of images. Several storage formats for pixel maps are used internally to GAME. To the user all images appear as a uniform class.

Images of native classes are stored in a compound format which uses a header made of parameters and tables to describe the data in the **datacube**. The datacube is the most primitive class of an image. Several images may be packed into one file and interrelated by HDB nodes. Images written in FITS format according to Wells et al. (1981) may be read on-line by the APs using GIFs. Therefore, data need not be preprocessed by a FITS reader.

An object of class **SIP** consists of a header and a pixel map. The header contains various information about the segmented pattern. The pixel map contains the picture of the pattern. A collection of SIPs may be stored as an object of class image. Information concerning all SIPs is stored in the header of this higher ranked object.

4.3 Session Environment

Data reduction takes place in a session environment (SE). An SE is represented as an instance of the respective class. An SE describes the capabilities of the objects participating in the data reduction session. One of these objects is the user, who may be a human being or an expert system. The SE registers states of the data reduction session in terms of parameters.

The SEs are ruled by a method called the **session monitor** (SM). The SM decides what action shall be taken, depending on its "knowledge" of the user's capabilities and the status of the session. Based on this information the SM initialises the event lists (see below) of the object classes to assure that no action, which is not in agreement with the capabilities of the user, is taken by any method.

SMs are not implemented as ordinary programs. A program called the administrator of session monitors (ASM) is driven by tables that define the functions and functionalities of the actual SMs. The definition of an SM is read from disk when a session is begun. The definition is selected by an entry in the SE. The user may change his SM. Individual communication methods can thus be installed for the users, where users may be humans, expert systems or programs of other types. An SM interfaces to the user via a communication method which may be a command language or menu or some graphic representation (icons). The SM supervises the processes initiated by the user and communicates results and status information to the user. The functional details of each SM can be varied over a wide range by its definition. Thus the ASM possesses properties of a user interface management system (UIMS; Green, 1985).

When the SE supports an expert system, messages to and from the system must be considered as data. An expert system may send instructions to the ADAS and interpret messages from the ADAS. The SM can function as an interface to the user of the expert system and as an interface between the expert system and the ADAS or even between several expert systems. A model of interaction between human beings, an image processing system and expert systems has been introduced by Goldberg et al. (1985). We found that running APs without SM, i.e. with operating system support only, in interactive mode is a matter of convenience. An SE with SM becomes a necessity as the user port as one steps towards an expert system, because it is the task of the SM to supply the appropriate working shell. The definition (file) of an SM may well be elaborated using an AI language such as Lisp.

The code of APs must supply adequate information to permit running the program in classical and AI mode. GAME uses source code processors to achieve correctness and completeness of APs. Source code processors are also used in IRAF (Image Reduction and Analysis Facility; Tody, 1987). It has been a goal of the design of GAME, that the algorithm is implemented in Fortran, while additional information is given in comment lines. A set of commands has been defined to mark respective comment lines. Thus a source code processor can automatically extract pieces of documentation from the source code. These pieces are stored in a database. They may be retrieved in arbitrary formats, corresponding to some definition. Thus the same database is used to print the manual and maintain on-line help support. The source code processor will also evaluate the GIF statements to generate respective entries in the SE.

4.4 Accessing Objects

When an AP is started by the user, it has to gain access to the objects. An AP can only connect to objects that are registered in the SE. An AP can never create an object from scratch. Therefore, the AP introduces itself to the SE at the very beginning (login). Thereafter, objects are activated from the SE by sending a message to a parameter of the SE. The message specifies what properties (class etc.) the new object is expected to exhibit. This specification is compared with the specification which is kept in the parameter. When both specifications match, the object is activated. If they differ, GAME tries to make the object look like expected. If no agreement can be reached, the operation fails.

In return, following the activation of an object, the AP receives a handle. Messages which are sent to the respective object must be attached to this handle. Common messages are concerned with reading, writing and clearing the data of an object.

Objects may react to a message with different methods depending on their internal status. The internal status is described in terms of an event list. The GAME event mechanism is a tool to alter the response of an object. The default goal of the methods implemented in GAME is to complete the requested operation. Each situation where the straightforward path to the completion of

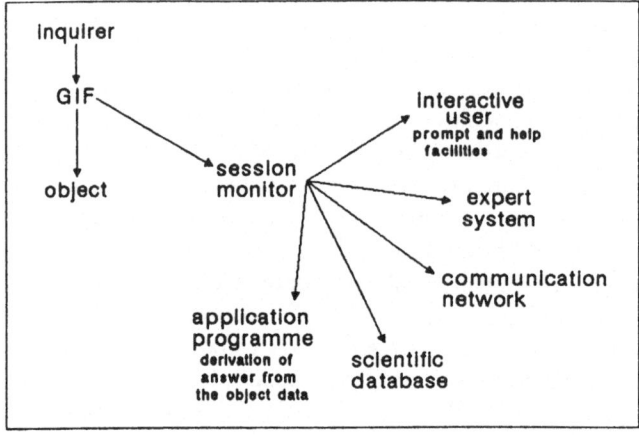

Fig. 4.1. Access paths to objects

the request is left, is called an event. Each event has a name. All events, whose names are listed in an event list, are trapped when they occur. If the trapped event has a negative character, the method completes with a return code that is the negative of the index of the event in the event list. If the trapped event has a positive character, additional operations are performed. APs may use the event mechanism with almost every call of a GIF and can thus react to the status of the SE at any time. This includes checks of the processing history of objects and of the user's environmental settings.

Interactive work is accomplished by reading or writing parameters from/to the SE or some other object. This is the only way for an AP to communicate to the user. Fig. 4.1 illustrates how a simple request to obtain a value can be satisfied. The simplest case is the presence of the value in the addressed object. The value is immediatly returned. If the requested number of values is not present, GAME can either report this fact to the AP or pass the request to the SM. The SM may distribute the request to many destinations depending on the the session status.

Objects are accessed using a few messenger interfaces. The Fortran calling syntax is

CALL GIF(OBJ, GCC, VALUE, RC)

where

CALL   is the FORTRAN CALL keyword,
GIF   is the name of the interface representing the message
OBJ   is the handle of the object the message is directed to,
GCC   is the GAME Command Chain,
VALUE is the location of the data,
RC   is the return code.

The GCC is a string which is used with most GIFs. It controls the access. At least it has to contain the name of the parameter to be read. Three succeeding sections may be specified.

The optional second section of the GCC contains a short comment on what the meaning of the parameter is. This comment is displayed (among other information) when the SM directs the request to the user in interactive mode.

The optional third section specifies a list of event names. The default goal of the access methods is to complete successfully by trying to put/get the values in/from a suitable location or even activate a different method to compute the values.

The fourth section contains the description of the object. When activating an object, it must at least specify the class of the object. Otherwise it may specify restrictions on, or modifications of, the object representation.

When in interactive mode and no communication method is installed with the SM, a primitive help function is activated. According to the user's choice the following prompts are issued:

> display of: default values,
> short prompt from section 2 of GCC,
> long explanation from help database,
> I/O-format.

The order in which these steps are taken depends on the user's status.

Table 4.2. Fixed command set.

command	action
.S	stop the AP unconditionally and immediately
.P	pause the AP and return to operating system command level
.C	cancel prompt (return to AP with fewer values than requested
.M	escape to session monitor
.L	display internal administrative tables (symbol queue)

The user is not limited to entering the value he is prompted for. He may also respond with commands to every prompt. A set of fixed commands (Table 4.2) aids in debugging APs and facilitates interactive communication. The AP may optionally install an additional set of commands, which can be used to efficiently specify standards or modify algorithms. Fig. 4.2 shows a sample program.

```
      PROGRAM MAKPSF
C Whenever columns 1 and 2 contain the characters 'C+' a new section of the on-line
C help information is collected. The kind of section is determind by the tag in
C column 3. Instead of a real specification the meaning of the column is explained
C below. A 'C-' terminates a section.
C+A   name of programmer
C+D   date of latest revision
C+P   description of purpose (a maximum of 2 lines)
C+I   shell instruction to activate this programme (if any)
C+C   general comments about this programme
C     ... as much as you want to write about the sense or nonsense of the programme
C+B   annotations to the bugs of the programme (if any)
C+S   long explanation for parameters
C     (This command may appear more than once in the programme.)
C     ... as much as you want to write about the sense or nonsense of the parameter
C-------------------------------------------------------------------------------
      IMPLICIT INTEGER*4  (A-Z)
      REAL*4     SEEING, SIZE
      DIMENSION  NAXIS(3)
C==============================================================================
C Introduce to environment:
      CALL  LOGIN ( ENV, 'MAKE_PSF', IST )
C
C Get the seeing for which the PSF shall be calculated:
      CALL  RKEYR ( ENV, 'SEEING;** at exposure time', SEEING, IST )
C     The seeing is expected to be supplied as run time parameter via the session
C     environment (SE). No help information is included in the on-line help library
C     for a possible prompt.
C
C Now get an image to work with:
      CALL  LOOKUP( PSF, 'SPECIMEN.4 ;;; IMAGE=2*14', NAXIS, NDIM )
C+S   SPECIMEN.4
C     A two-dimensional image (data as 4 byte integers) containing a well exposed
C     star. The image of the star is used to determine the PSF.
C-
C Determine what part of the image to take into account:
C     Set commands and read parameter with event COMMand trapped.
      CALL  SETCOM( 'GOOD,BAD' )
      CALL  RKEYR ( PSF, 'SIZE;size of specimen;COMM,', SIZE, IST )
C+S   SIZE
C     The size of the specimen is missing in the header. Specify it here or
C     calculate it according to the seeing. If you enter
C     GOOD   the size equals    5 * seeing
C     BAD    the size equals   20 * seeing
C-
C  -- Size defined by commands?
      IF      ( IST .EQ. -1 ) THEN
C     Yes, first event (=one and only) was trapped! Get index of command:
C               CMD = GETCOM( 0 )
      IF      ( CMD .EQ. 1 ) THEN
C     - Install default for GOOD seeing:
         SIZE =  5 * SEEING
      ELSE IF ( CMD .EQ. 2 ) THEN
C     - Install default for BAD seeing:
         SIZE = 20 * SEEING
      END IF
C
C  -- No values for input or any severe errors ?
      ELSE IF ( IST .NE.  1 ) THEN
C     - Inform user.
         CALL PUTMSG('TAKECARE')
         ... error handling ...
      END IF
C
      ... algorithm ...
C
      CALL  LOGOUT( PRG, 'DONE', IST )
      END
```

Fig. 4.2. Sample program. Only rudimentary declarations are used and return codes (last argument) are not checked in every case to keep this figure small enough.

4.5 FLOPP: A Program for Automated Extraction of Image Parameters

Fig. 4.3 is a graphic representation of the reduction process for a direct Schmidt plate. The reduction starts with the scanning and on-line segmentation (search for objects) of the plate. It leads to an image-type file containing global plate parameters in the main file header and the segmented astronomical objects in the form of SIPs. The next step is feature evaluation which yields a list of object parameters. These are used to perform the object classification, which results in a list of stars, galaxies and rejected objects. In the following, a more detailed description of the process of feature evaluation (program "FLOPP") is given.

The task of the program FLOPP is to extract information from segmented objects. The isolated objects are rectangular segments of the digitised astronomical plate which are stored as SIPs. Each SIP contains a galaxy or stellar image. All SIPs of one plate are collected as one image.

Parameters related to a single segment (local sky density, local plate noise) are kept in the SIP. Parameters concerning the whole astronomical plate (characteristic curve, astrometric parameters) are stored in the image header.

FLOPP needs the following information to process the segmented objects:

(a) parameters of the photographic plate, such as
   – characteristic curve (as a function of x and y),
   – sky density (as a function of x and y),
   – plate noise (as a function of x and y),
   – point spread function of stellar images (as a function of x and y);
(b) the image data of the objects;
(c) object parameters that have been computed from the images, such as
   – central object coordinates (center of mass),
   – central intensity,
   – object radius (effective radius),
   – size (number of pixels covered),
   – apparent magnitude,
   – ellipticity,
   – object orientation;
(d) processing history of the data:
   – reduction algorithms and parameters at different stages of the reduction process;
(e) information about object vicinity (location and brightness of neighbouring images);
(f) lists and identifications of astrometric/photometric standards, astrometric transformation parameters.

The processing of a SIP is controlled by a message string. The methods may be preprocessing operations (filtering, thresholding) or the determination of various image parameters (determination of central image density, moments).

The message string is established by an interactive routine which helps the user to build the string from menus. Explanations of methods are given on

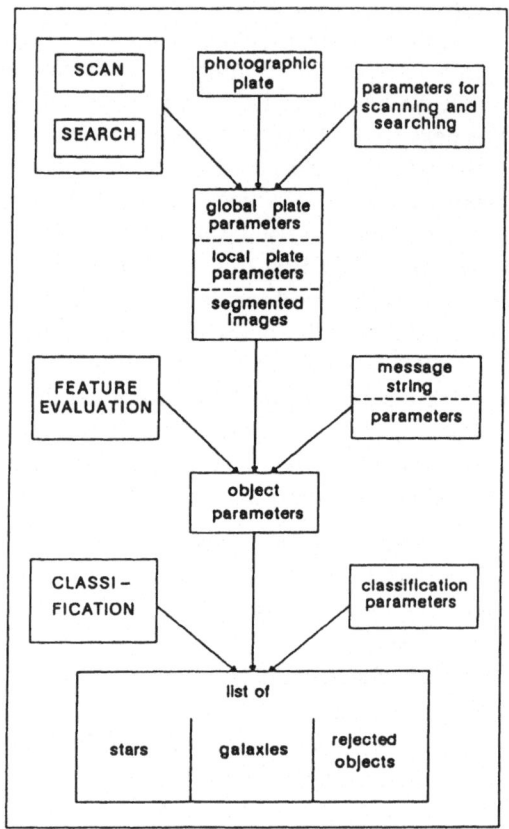

Fig. 4.3. Processing of the direct photographic plates.

request. Warnings are displayed, including suggestions for correction, if incompatible sequences are chosen.

4.6 METASC: A Program for Automated Classification of Astronomical Spectra

METASC consists of several methods (Table 4.3) for the reduction and classification of astronomical spectra. The isolated spectra are represented as SIPs.

Table 4.3. Some methods for the reduction and classification of astronomical spectra applied in the MRSP.

methods
spectrophotometry
spectrum rectification
calculation of colours, spectral moments
correlations, least squares fits between spectral profiles
ratios and differences between spectral profiles
detection and identification of spectral features
Bayes classifiers, fuzzy set classifiers, NN-classifiers.

The methods are arranged in a hierarchical structure that implicitly contains their interdependencies.

The user browses through a menu of messages and selects the messages necessary to perform a specific task. The menu is arranged in a hierarchical structure which is similar to the arrangement of the methods. A collection of several messages is called a message string. Message strings have been developed for the classification of stellar spectra (Schuecker et al., 1986), for the identification of quasars (Gericke, 1988), and for galaxy redshift measurement (Schuecker, 1988).

The creation of a message string is supervised by a help system which

– inserts messages necessary for correct execution to the user–specified message string,
– detects/corrects inconsistent choices of messages,
– gives hints of possible improvements for the actual string.

The help system consists of several hundres "if-then" rules of the form

> Rule (n) IF (message k present) then
>   include message i
>   ...
>   include message j

For general messages k, corresponding to general methods, intermediate and/or specific messages i,j are included whereas specific messages k include intermediate and/or general messages i,j. The rules are ordered in a hierarchical structure, similar to the hierarchical structure of the methods.

Assume the user selected the message "median filtering", which fires Rule (n):

```
OBJECT FILTER : Read spectrum
                Discard non-stellar spectra
                Discard overlapping spectra
QUALITY FILTER: Local sky background/noise
                Discriminate spectra by S/N
SCENE ANALYSIS: Transform density to intensity
                Spectrum profile        :        Point spread function
                                                  Max.Likelihood method
                Spectrum filtering       :        Fuzzy set rectification
                                                  ... in highpass ...
                Wavelength calibration :          Position of emulsion
                                                              cutoff
SHAPE         : Global feature          :        Correlation function
                Local  feature          :        Derivation of SHAPE
DESCRIPTION   : Classification          :        Fuzzy set classifier
```

Fig. 4.4a. Message string for the classification of stellar spectra.

```
SCENE ANALYSIS: Spectrum filtering       :        Differential filtering
SHAPE         : Local feature            :        Gauss fit to CaII break
                                                  (x,y) position of break
```

Fig. 4.4b. Message string for the measurement of the continuum break positions in stellar spectra. Take spectrum tracing after maximum likelihood calculation of the spectral profile as processed by the message string given in Fig. 4.4a.

> Rule (n) IF ('Median filtering') then
>   include: 'filter spectral profiles'
>   include: 'do Scene Analysis'

The presence of the second message fires Rule (m):

> Rule (m) IF ('do Scene Analysis') then
>   include: 'notify: calibration changes'
>   include: 'calculate object profiles'

which in turn fires Rule (h):

> Rule (h) IF ('calculate object profiles') then
>   include: 'load object data'

Additionally, Rule (m) notifies the user about possible modifications regarding the calibration of the object data, the correction of the object intensities for atmospheric and/or detector effects, etc.

Before METASC starts to process astronomical spectra a message string is chosen and copied into the short-term memory STM (see below). At present, METASC is able to work with a maximum of two message strings simultaneously. The application of several message strings reduces I/O and CPU times, because methods, common to both message strings, do not have to be applied several times for the same object.

Examples for message strings are given in Figs. 4.4a and 4.4b. Message string (a) identifies G-type stellar spectra and string (b) measures the position of the CaII-feature of classified G-type stars for astrometric purposes.

The parameters used in METASC can be divided into program-technical parameters (long-term memory LTM) and training parameters (TP). The LTM is determined during the program development phase and should never be changed for a specific survey task. LTM parameters are e.g. parameters organizing the short-term memory STM (see below), the data I/O formats, storage etc.

TPs are determined during the training phase and might change (from plate to plate). TPs are parameters describing the feature spaces for various classification schemes, spectral calibrations etc. External standards (prototypes), chosen by the user, are used for training. In most cases the number of prototypes is smaller than necessary for the reliable calculation of the feature spaces. Therefore, spectra similar to the prototypes are identified automatically on the actual field and are used as second-order prototypes. In METASC about 1,000 second-order prototypes are needed for the calibration of the wavelength reference point (Tucholke et al., 1989). More details about the training procedures applied in the MRSP are given by Schuecker et al. (1989a).

After processing of the direct plate and the spectrum segmentation of the objective prism plate, the individual spectra (SIPs) are processed by METASC. The LTM, the TPs and the actual message strings are stored in METASC in the short-term memory (STM). The STM insures that each method can use the results from all other methods by simulating a "blackboard", from which the individual methods get their input and on which they leave their output. The input/output domains in the STM are organised by the LTM.

4.7 Concluding Remarks

We have presented an outline of the software implemented for the MRSP. Automated abstraction of several million astronomical objects relies on a high standard for the software. AI level software has been implemented in Fortran. Examples for process and database management and for object processing and classification are given. From the point of view of object oriented programming, most existing APs can be regarded as implementations of methods. Up to now, about two million astronomical images have been processed by the system described in this paper. To demonstrate a result Fig. 4.5 shows a map of the two-dimensional galaxy distribution obtained from the reduction of three Schmidt plates. A prominent supercluster can be seen in the centre of the map. Detailed results are given by Horstmann (1988; 1989) and Schuecker et al. (1989b).

Acknowledgements. This work is partially supported by the Deutsche Forschungsgemeinschaft. For their assistance in implementing the software, we have to thank R. Budell and S. Grefen.

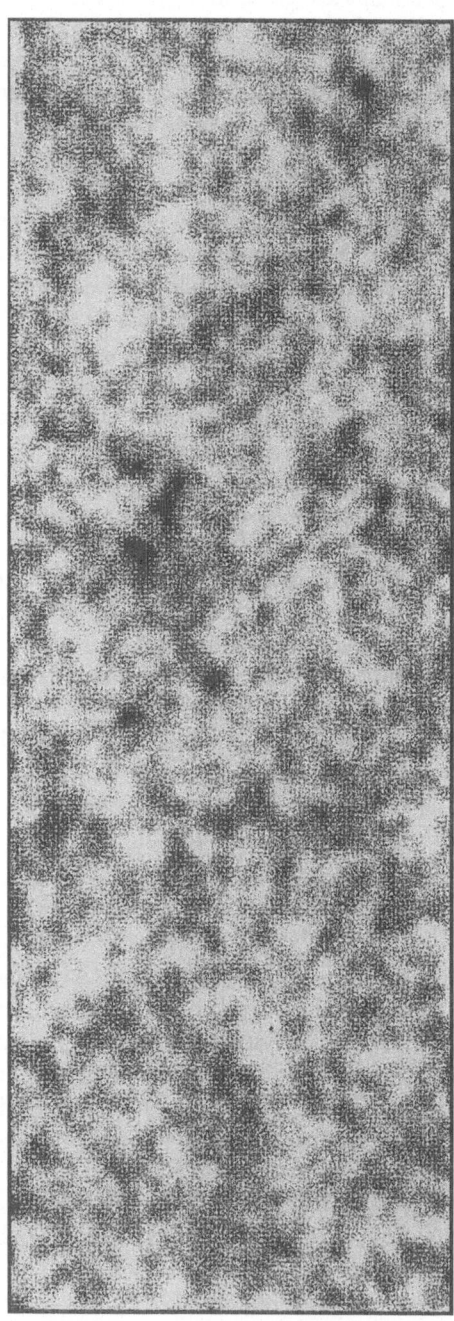

Fig. 4.5. Map of two-dimensional galaxy distribution from fields 410, 411 and 412 (top to bottom) of the ESO/SRC survey. These fields cover about 70 square degrees close to the south galactic pole. Limiting magnitude is about 21 mag. More than 80,000 galaxies have been included. Dark regions indicate a high number density.

References

1. Gericke, V. (1988), "A search for homogeneous samples of quasars", *Large-Scale Structures in the Universe: Observational and Analytical Methods*, W.C. Seitter, H. Duerbeck and M. Tacke, M. (eds.), Springer-Verlag, Heidelberg, 235–245

2. Goldberg, A. and Robson, D. (1983), *SMALLTALK 80: The Language and its Implementation*, Addison-Wesley, Reading, MA.

3. Goldberg, M., Goodenough, D.G., Alvo, M. and Karam, G.M. (1985), "A hierarchical expert system for updating forestry maps with Landsat data", *Proceedings of the IEEE*, **73**, 1054–1063.

4. Green, M. (1985), "Report on dialogue specification tools", *Workshop on User Interface Management Systems*, Proceedings, G.E. Pfaff (ed.), Springer-Verlag, Berlin, 9–20.

5. Horstmann, H. (1988), "Automated analysis of galaxy clustering" *Large-Scale Structures in the Universe: Observational and Analytical Methods*, W.C. Seitter, H. Duerbeck and M. Tacke, M. (eds.), Springer-Verlag, Heidelberg, 111–121.

6. Horstmann, H. (1989), "A statistical approach to morphological classification using apparent ellipticities of galaxy images", *Proceedings of the Workshop on Large-Scale Structures in the Universe*, Publications of the Observatory of Paris/Meudon, in press.

7. Schuecker, P. (1988), "The Muenster Redshift Project", *Large-Scale Structures in the Universe: Observational and Analytical Methods*, W.C. Seitter, H. Duerbeck and M. Tacke, M. (eds.), Springer-Verlag, Heidelberg, 142–159.

8. Schuecker, P., Horstmann, H. and Volkmer, C.C. (1986), "Automatic extraction of very low-dispersion spectra", *Data Analysis in Astronomy II*, V Di Gesù, P. Crane, J.H. Friedman and S. Levialdi, S. (eds.), Plenum Press, New York, 109–116.

9. Schuecker, P., Horstmann, H. and Teuber, D. (1989a), "Automated classification applied to wide-angle astronomical photographs", *Klassifikation und Ordnung*, R. Wille (ed.), in press.

10. Schuecker, P., Ott, H.-A., Horstmann, H., Gericke, V. and Seitter, W. (1989b), "Observations of cosmic fluctuations", *Proceedings of the Third ESO-Cern Conference on Cosmology*, Kluwer, Dordrecht, in press.

11. Shaw, M. (1984), "The impact of modelling and abstraction concerns on modern programming languages", *On Conceptual Modelling*, M.L. Brodie, J. Myolopoulos and J.W. Schmidt (eds.), Springer-Verlag, New York, 49.

12. Teuber, D. (1985), "An astronomical data analyzing monitor", *Data Analysis in Astronomy*, V Di Gesù, P. Crane, J.H. Friedman and S. Levialdi, S. (eds.), Plenum Press, New York, 235–239.

13. Teuber, D. (1988a), "The Generic Applications and Monitors Environment", *Astronomy from Large Databases: Scientific Objectives and Methodological Approaches*, F. Murtagh and A. Heck (eds.), European Southern Observatory, Garching bei München, 261–266.

14. Teuber, D. (1988b), "The hardware and software support for the MRSP", *Large-Scale Structures in the Universe: Observational and Analytical Methods*, W.C. Seitter, H. Duerbeck and M. Tacke, M. (eds.), Springer-Verlag, Heidelberg, 323–330.

15. Teuber, D. (1989), "GAME System Manual", Version 3.1, in preparation.

16. Tody, D. (1987), "The IRAF data reduction and analysis system", in *IRAF System Handbook* 3A, NOAO.

17. Tucholke, H.-J., Schuecker, P., Horstmann, H. and Seitter, W. (1989). "Astrometric determination of the reference point for wavelength measurements on low-dispersion objective prism plates", In IAU Colloquium 100, *Fundamentals of Astrometry and Celestial Mechanics*, H. Eichhorn and I. Pakvor (eds.), in press.

18. Wells, D.C., Greisen, E.W. and Harten, R.H. (1981), "FITS: a flexible image transport system", *Astronomy and Astrophysics Supplement Series*, **44**, 363–370.

5 Distributed Point-Pattern Matching

George R. Cross (1) and Rittick Gupta (2)

(1) Department of Computer Science
Washington State University
Pullman, WA 99164-1210
USA

(2) Department of Computer Science
Louisiana State University
Baton Rouge, LA 70803-4020
USA

5.1 Introduction

We continue our previous effort (Dodhiawala and Cross, 1986) to develop a
system to track the linear movement of cosmic ray particles in three dimensional space. We adapt techniques of distributed problem solving (Decker, 1987)
and knowledge-based systems (Cross, 1986) to problems in astrophysical image
processing. Distributed problem solving is the cooperative solution of a single
problem by a group of decentralized and loosely coupled knowledge sources. Mutual sharing of information is necessary because none of the knowledge sources
is independently capable of arriving at the solution from the incomplete, and
possibly inconsistent, information that is locally available to it.

5.2 JACEE

The purpose of the Japanese-American Cooperative Emulsion Experiment (JAC-EE) is to study cosmic rays (Huggett et al., 1981). A sandwich of photographic emulsion plates is sent aloft on a flight using a balloon. Cosmic rays strike the sandwich at a target made of lead and emit secondary particles. The particles travel through a stack of plates, leaving behind a trace in the form of a spot on the emulsion. The distance travelled depends on the energy of the particles. We need to determine the path of the particle through the plates. Computerization is needed because the manual process is error-prone, time-consuming, and tedious.

Viewed from the top, the plates look like collections of points. The points must be correlated to determine the location of the *vertex* of the cosmic ray event. In addition, the successive frames may have measurement errors, points missing, and spurious noise points added. Furthermore, there may be secondary events, i.e. subtrees of points.

5.3 Experimental Apparatus

The apparatus includes a *primary charge detector* which is used for determination of charge of each primary particle. The *target* consists of emulsion plates interleaved with Z-material to maximize the nuclear interaction. *Spacers* diverge closely collimated gamma rays. The *calorimeter* is a set of lead sheets interleaved with X-ray films and thin emulsions. Events are detected by observation of dark spots produced by cascades in the X-ray film. The coordinates of dark spots in the X-ray films are mapped directly onto the adjacent emulsion plates, and the cascades are located using a microscope.

5.4 Distributed Problem Solving

Distributed problem solving is the cooperative solution of problems by a decentralized and loosely coupled collection of knowledge sources, located in a number of distinct processor nodes. The two forms of cooperation in distributed problem solving are task sharing and result sharing. The advantages of distributed problem solving include: speed, reliability, the ability to handle applications with natural spatial distribution, and the ability to tolerate uncertain data and knowledge.

One of the techniques of distributed problem solving is a blackboard architecture (Nii, 1986b). Blackboards have been found to be useful in problems where there is a natural transformation from signals to symbolic information like speech understanding and acoustic signal processing (Nii, 1986a). The *blackboard data structure* is used in the same way that scientists use a blackboard in a room when they work together to solve a problem. The blackboard contains hypotheses about the solution of the problem. At any given time, there may be competing and even contradictory propositions on the blackboard. Information from the signal sources is processed and integrated into the blackboard; eventually unlikely hypotheses are discarded. The correct ones remáin as the problem solution.

The raw data from the environment is not itself used to update the blackboard. This is instead done by independent *knowledge sources* that filter and interpret data according to their specialized knowledge base. Although the knowledge sources in the system are independent, they cooperate to find a solution.

Knowledge sources are scheduled for execution by a *blackboard monitor*. The system is said to be *data-directed* since the knowledge sources are scheduled for execution when their *triggering pattern* is matched. The blackboard monitor executes a *precondition procedure* for each knowledge source to determine if it should be scheduled for execution. Scheduling is necessary since access to the blackboard must be serialized.

We used a "domain-independent" system called HEARSAY-III (Balzer et al., 1980; Erman et al., 1981) in the original version of this research (Dodhiawala and Cross, 1986), but we have subsequently abandoned it because it was very slow and the only computer/software combination that it ran on is no longer available to us (DEC VAX/VAX Interlisp).

5.5 The Computational Process

Our previous work (Dodhiawala and Cross, 1986) exploited the minimal spanning tree and subgraph isomorphisms under the aegis of the HEARSAY-III software package. We have re-implemented the entire problem, including more knowledge of the physics of the problem, using a mixture of the computer languages Lisp and C. Multiple processes are used to simulate distributed processing.

5.5.1 Flavors

We use the object-oriented data structures called Flavors (Moon, 1986). Flavors are the intellectual precursor to the more current Common Lisp Object System (CLOS; Bobrow, 1988). In our use of Flavors, we have three levels of objects: Plates, Points, and Tracks. Each plate has been represented as a Flavor with the following slots:

Points: Contains the coordinates of all the points in the plate.
Cor-Points: Contains the set of points in adjacent plate corresponding to a point and the associated scores.
Missing-Points: List of the coordinates of the generated points.
Noise: List of noise points.

Each track has been represented as a Flavor with the following slots:

List: An association list of plate name and point number forming the track.
Focus: Coordinates of the origin or vertex of the track.
Electron/Proton: Type of particle, but only one of them is set to true.

Points are represented as a record type with the following slots:

Possible: List of possible corresponding points.

Tried: List of points which have been tried from the possible list.

XX: Extrapolated X coordinate.

YY: Extrapolated Y coordinate.

Name: Track to which point was assigned.

Dist: Distance of point from the extrapolated point.

Index: Number assigned to point.

5.5.2 Algorithms

5.5.2.1 Initial Processing. We first find a common axis for each adjacent plate. We then find a set of possible corresponding points for each point in a plate. Then we compute the first approximation of a vertex. We next attempt to extend tracks. Finally, we recompute the vertex. At this stage, we identify the type of track (i.e. electron or proton), and then reassign points in secondary tracks.

We need to explain how we get the initial approximation to the tracks. Let

$$P = \{p_1, p_2, p_3, ..., p_m\}$$

and

$$Q = \{q_1, q_2, q_3, ..., q_n\}$$

be two patterns of points on different plates. A figure of merit is assigned to each pair (p_i, q_j) according to how closely other pairs (p_n, q_k) match when p_i is mapped to q_j. The points with maximum similarity in the first two plates are joined. The intersection of these lines is the first approximation of the vertex.

5.5.2.2 Matching Algorithms. Three methods have been developed for solving this problem. We explain each in detail.

5.5.2.3 Relaxation I. A relaxation-like technique has been used to match the patterns on each adjacent plate. No initial orientation of the plates is assumed. Multiple processes are used for this phase of the problem. Because of limitations of the Unix system (i.e. process communication using sockets is limited to characters only), disc files are used for interprocess communication. This method finds the best match between two patterns by assigning a point on a pattern to one point in the other pattern. The other points are then assigned on the basis of distance on the x-y plane. A score is computed, which indicates how well the patterns match. This process is repeated for all points in the second pattern. The assignments with a maximum score represent the best possible match between the two patterns and are retained for further processing. An initial approximation of the vertex is computed by joining the corresponding points on the first two plates. Tracks are then formed by joining the corresponding points on the plates. Evidence regarding the type of particle is computed based on the distance of the extrapolated point from the actual point. Secondary effects are detected when a new track is formed on a plate other than the first plate. Although efficient, this method has problems assigning points to tracks in a high density cascade.

5.5.2.4. Relaxation II. This method also uses a relaxation-like technique. However, instead of a single point, a set of points is assigned to each point in an adjacent plate. The first approximation for the vertex is computed by joining points which correspond best. Some of these intersection points are outliers and are rejected; the average is used as the first approximation. The corresponding points on the plates which form a track is found by extrapolating the track and selecting the point closest to the extrapolated point. This assignment of points to tracks can also be viewed as a truth maintenance problem (Doyle, 1979). Initially points are assigned based on the justification of being closest to the track. These assignments are then updated if and when conflicts are found or the justifications are violated. The vertex or the focus of expansion is refined each time points in a plate are assigned to tracks. Formation of cascades are detected by tracks being formed at plates other than the first plate. The set of possible points mentioned earlier is used to find the vertex of the cascade. Tracks formed in a cascade are very dense near the origin and they spread out, away from the origin. The possibility of assigning a point to a wrong track is a danger close to the origin because of the high density. An evidence-based technique has been used to resolve this problem. After the tracks have been formed, evidence regarding the type of particle is computed. The function used is similar to the error term in the least squares method. The points near the origin are reassigned based on this evidence. Light particles and scatter are taken into account for computing the extrapolated point, in the assignment method mentioned earlier. This procedure may be repeated several times.

5.5.2.5 Blackboard Version. This method is similar to the method mentioned above. The control structure has been designed to simulate distributed problem solving. We have two blackboards, a Scheduling Blackboard and a Fact Blackboard. The Scheduling Blackboard is used to control the interpretation activities while the Fact Blackboard contains global and local data. Extendable tracks are scheduled by the scheduler on the scheduling blackboard.

Tracks have been implemented as a Flavor with Methods being used for communication. Since tracks are objects they can inform neighboring tracks of the success of an assignment attempt. The message sent to a track is a list of selected points and distance from the extrapolated point. The message indicates how close the extrapolated track is to the assigned point. Tracks receiving a message are then rescheduled by the scheduler. The intersection of tracks is used to obtain a better estimate of vertex. A line is fitted through points in a track and an error term is used to determine type of track.

Conflicts arise when the same points are assigned to two or more tracks. Conflicts are resolved by passing messages. Conflicts are resolved by tracks reassigning points till each point is assigned to one track only. Evidence gathered regarding the type of track is used to reassign points in a secondary track.

5.6 Experimental Results

5.6.1 Simulation of Points

Points are generated on plates by using a random number generator. Plates are of 100 by 100 elements and with a resolution of 0.05 units. Twenty points are generated on the first plate. The initial vertex is at $(50, 50, -20)$. There are on the average thirty points on a plate. The following is the distribution of points:

1. 5% noise
2. 10% extra points.
3. 5% of the points cause secondary phenomena.
4. 45% of the points are due to heavy particles (protons), i.e. there is no scattering effect.
5. 50% of the points scatter, i.e. they are electrons or positrons. On the average scattering is limited to 60 degrees between two plates.

5.6.2 Results

The results of 5 sample runs are shown in Tables 5.1 through 5.5 below. Overall, matching results are good, with some difficulties noted in the secondary particles.

5.7 Future Directions

At present we are re-implementing the system on a Sequent Balance parallel processor (cf. Hwang, 1987). We would like to be able to handle multiple tracks in each cascade and make use the angle of incidence of each track at each plate.

Table 5.1. Experimental results: sample 1.

Plate#	Total	Extra	Missing	Light	Heavy	Primary	Secondary
1	20	2	0	9	9	18	0
2	21	2	1	9	9	17	1
3	24	2	0	13	5	17	5
4	26	2	1	14	9	20	3
5	31	2	2	18	9	21	6

Results:

All primary particles were correctly matched.
70% of the secondary particles were correctly matched.
All missing points were correctly detected.

Table 5.2. Experimental results: sample 2.

Plate#	Total	Extra	Missing	Light	Heavy	Primary	Secondary
1	20	1	0	9	10	19	0
2	21	1	3	11	9	19	1
3	34	1	0	24	9	15	18
4	34	1	3	21	9	30	0
5	39	1	8	9	20	24	5

Results:

98% of the primary particles were correctly matched.
50% of the secondary particles were correctly matched.
All missing points were correctly detected.

Table 5.3. Experimental results: sample 3.

Plate#	Total	Extra	Missing	Light	Heavy	Primary	Secondary
1	20	1	0	11	8	19	0
2	26	1	2	15	8	15	8
3	35	1	2	26	8	23	11
4	39	1	3	30	8	33	8
5	45	1	5	36	8	38	6

Results:

95% primary particles were correctly matched.
65% of the secondary particles were correctly matched.
All missing points were correctly detected.

Table 5.4. Experimental results: sample 4.

Plate#	Total	Extra	Missing	Light	Heavy	Primary	Secondary
1	20	1	0	2	17	19	0
2	27	1	2	9	17	17	9
3	29	1	2	10	17	24	3
4	36	1	1	18	17	26	9
5	36	1	1	18	17	35	0

Results:

All primary particles were correctly matched.
80% of the secondary particles were correctly matched.
All missing points were correctly detected.

Table 5.5. Experimental results: sample 5.

Plate#	Total	Extra	Missing	Light	Heavy	Primary	Secondary
1	20	1	0	10	9	19	0
2	26	1	1	16	9	16	6
3	33	1	1	23	9	22	10
4	33	1	2	23	9	32	0
5	38	1	4	28	9	30	7

Results:

All primary particles were correctly matched.
60% of the secondary particles were correctly matched.
All missing points were correctly detected.

References

1. Balzer, R., Erman, L.D., London, P. and Williams, C. (1980), "HEARSAY-III: a domain-independent framework for expert systems", in *Proceedings of the First Annual Conference on Artificial Intelligence*, 108–110.
2. Bobrow, D.G. (1988), "The Common Lisp Object System: an example of integrating programming paradigms", *Exploring Artificial Intelligence*, H.E. Shrobe (ed.), Morgan Kaufmann, San Mateo, CA, 619–640.
3. Cross, G.R. (1986), "Tools for constructing knowledge-based systems", *Optical Engineering*, **25**, 436–444.
4. Decker, K.S. (1987), "Distributed problem solving techniques: a survey", *IEEE Transactions on Systems, Man, and Cybernetics*, **SMC-17**, 729–740.
5. Dodhiawala, R.T. and Cross, G.R. (1986), "Analysis of cosmic ray tracks using distributed problem solving", *Pattern Recognition Letters*, **4**, 471–476.
6. Doyle, J. (1979), "A truth maintenance system", *Artificial Intelligence*, **12**, 231–272.
7. Erman, L.D., London, P.E. and Fickas, S.F. (1981), "The design and an example use of HEARSAY-III", *Proceedings of the International Joint Conference on Artificial Intelligence*, 409–415.
8. Huggett, R.W., Hunter, S.D., Jones, W.V., Takahashi, Y., Ogata, T., Saito, T., Holynski, R., Jurak, A., Wolter, W., Wosiek, B., Dake, S., Fuki, M., Tominaga, T., Friedlander, E.M., Hackman, H.H., Parnell, T.A., Miyamura, O., Gregory, J.C., Burnett, T.H., Lord, J.J., Wilkes, R.J., Hayashi, T., Iwai, J. and Tabuki, T. (1981), "Japanese-American Cooperative Emulsion Experiment", *17th ICRC Conference Papers*, Vol. 8, CEN Saclay, Paris, 80 pp.
9. Hwang, K. (1987), "Advanced parallel processing with supercomputer architectures", *Proceedings of the IEEE*, **75**, 1348–1379.

10. Moon, D.A. (1986), "Object oriented programming with Flavors", *OOPSLA '86 Conference Proceedings*, N. Meyrowitz (ed.), ACM, New York, 1–8.

11. Nii, H.P. (1986a), "The blackboard model of problem solving and the evolution of blackboard architectures", *The AI Magazine*, **7**, 38–53.

12. Nii, H.P. (1986b), "Blackboard application systems and a knowledge engineering perspective", *The AI Magazine*, **7**, 82–107.

6 Decision Problems in the Search for Periodicities in Gamma-Ray Astronomy. How Can A.I. Help?

M.C. Maccarone and R. Buccheri
Istituto di Fisica Cosmica ed Applicazioni dell'Informatica, C.N.R.
Via Mariano Stabile 172
I-90139 Palermo
Italy

6.1 Introduction

The detection of periodic signals from celestial sources of γ radiation is a very important step for understanding their astrophysical properties. Great effort has therefore been undertaken by several γ-ray astronomy groups for the development of suitable analysis packages. In spite of this, the analysis procedures currently used, both in satellite and ground based γ-ray astronomy, differ in many respects which makes reciprocal understanding of results difficult.

The various techniques in use have been discussed by De Jager (1987) and more recent developments have been published in Buccheri and De Jager (1988). In this paper we want to reiterate the evidence (already pointed out by Maccarone and Buccheri, 1988, 1989) that one of the major problems of searching for periodicities consists of the need for a number of "subjective" decisions. It is a matter of fact that, due to lack of knowledge of the properties of the signal under analysis (period, harmonic content, etc.) different techniques may be applied, each of them being able to furnish a solution to different problems. The actual choice depends on practical circumstances like the availability of working software packages, knowledge about similar analyses, statistical considerations about the data and so on. In general, these circumstances cannot be described formally with the result that any choice will have a certain degree of subjectivity.

We aim to investigate whether artificial intelligence may help to reduce the level of subjectivity of the decisions taking into account what is available concerning both data analysis techniques in γ-ray astronomy and artificial intelligence facilities.

6.2 Basic Problems in Periodicity Searches

The main steps to follow in the analysis of photon arrival times for the detection of periodic signals are illustrated in the flow chart of Fig. 6.1, part of which is reproduced from Maccarone and Buccheri (1988). Each of the rectangular boxes is relative to a set of standard processing techniques; the choice of the one to use in the actual case is made in the round boxes which therefore represent the nodes of the analysis where the decisions have to be taken. Most of the discussion on the techniques in use and the decisions to take can be found in the papers quoted in the introduction and we will refer to them for details. Here we want just to list again the decisions needed in order to link them with suitable artificial intelligence tools.

The decision nodes are as follows:

- which selection criteria to apply to the raw data. Different selection criteria will give rise to different final sets of arrival times which might have different statistical properties. As an example we can think of the choice of the energy range or time interval or photon coordinates around the source direction. The decision should be made as a first step of the analysis since repeating it with different selection criteria may significantly affect the final outcome. This point was not discussed in the previous papers where a fixed *a priori* selection was implied.

- whether to use the Fast Fourier Transform (FFT) or folding algorithms according to the size of the data sample. For large data samples, FFT techniques may be desirable in order to achieve affordable computation time for the analysis.

- which interval of periods to investigate. This depends on how good our knowledge of the expected period is. For purely exploratory cases this is a very crucial point since the sensitivity of the search decreases with the increase of the interval of investigation (see Fig. 6.2). For weak signals, practically always the case in γ-ray astronomy, it may mean that the sensitivity drops below the visibility threshold for sufficiently large intervals. Therefore a compromise has to be found by defining an acceptably small period interval which includes most of the expected physical properties.

- how to estimate independent trials. This is a direct consequence of the choices made in the first two points. Since in general a clear enumeration of the trials is not possible (due, for example, to oversampling or to the application of more than one selection criterion) the estimate contains usually a certain degree of subjectivity.

- how to define the significance level. This generally depends on the personal taste of the analyst, some being satisfied by reaching, say, a chance occurrence probability of 0.01 while others requiring much lower threshold values by

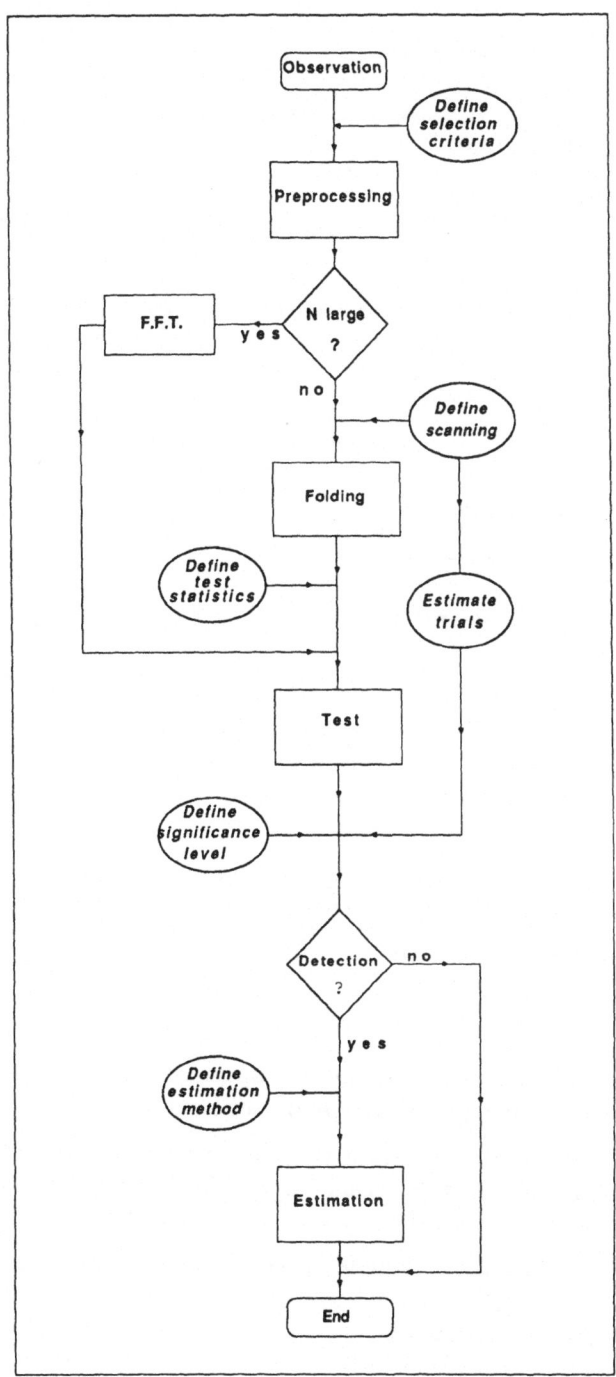

Fig. 6.1. General flow of operations in the search for periodicity in γ-ray astronomy.

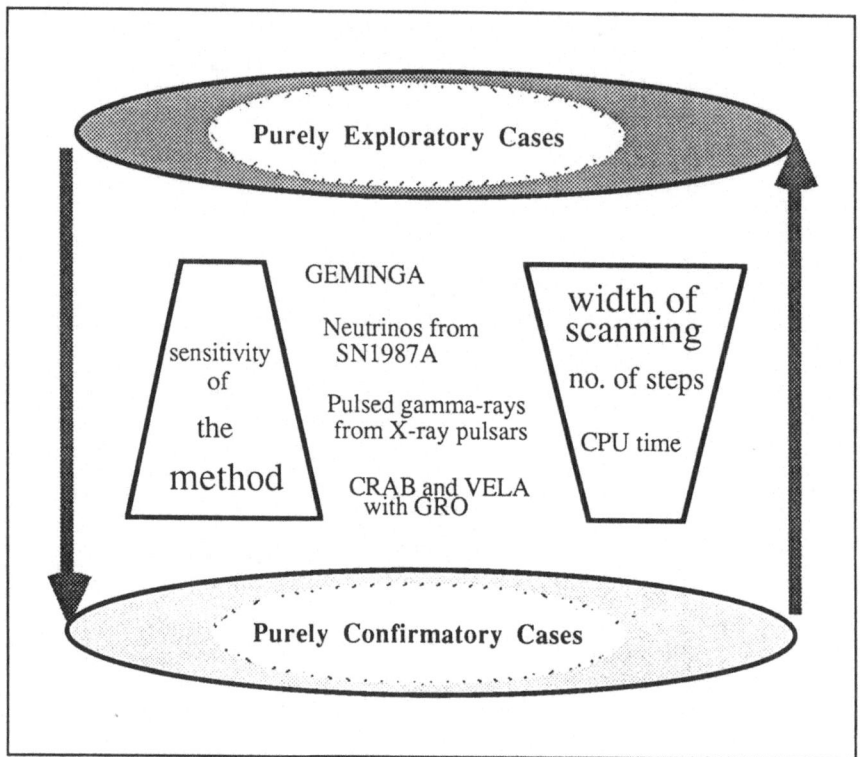

Fig. 6.2. Sensitivity of the search versus the width of the scanning interval. Some examples are illustrated going from purely exploratory cases (periodicity search in the case of the γ-ray source Geminga) to purely confirmatory cases (Crab and Vela pulsars to be confirmed by the Gamma Ray Observatory, GRO).

considering the unknown number of optimizations almost always present and the possible presence of systematic periodicities intrinsic to the data.

- how to define the test statistic. This decision implies a hypothesis on the harmonic content of the signal searched for and contains therefore some level of subjectivity especially in completely exploratory cases.

- how to choose the technique for estimating the characteristics of the light curve. Once the signal has been detected one has to chose whether to focus attention on small or on large scale structures of the signal. Different techniques are available in the two cases so again a decision is needed.

6.3 Towards a Decision Support System

The search for periodicities in γ-ray astronomy may thus be considered as a problem which needs several kinds of knowledge and which is is also related to databases which evolve as new information is introduced.

Therefore it is correct to think that artificial intelligence tools may be employed in order to define a support system. This system would be capable of assisting the user in the decision making phase proper to the data analysis process, especially in the selection of algorithms and methods.

To define the features of such a system, it is important to take into account several statements:

- The system must be linked with different, integrated environments.

It may be necessary to derive parameters and statistical considerations from the same actual data (for example, to derive the probability distribution related to the test to be used), by performing on them algorithms and procedures already implemented in conventional data analysis environments and which therefore only need to be accessed.

- The system must be linked with external databases.

Different information, directly or indirectly related to the actual data to be analyzed, can be contained in astronomical databases, numerical or bibliographical. These databases can be accessed by their query languages; so the system needs appropriate interfaces in natural language or similar between the user and the query languages of the databases considered.

- The system needs to deal with different types of knowledge.

The various kinds of knowledge (objects, concepts, definitions, relations, strategies, heuristics, and so on) are related to the way they are to be used. Utilization of a given kind of knowledge depends on objectives such as acquisition and learning of new facts, retrieval of facts concerning a given situation, and reasoning on what the system knows in order to determine what it is to do. The acquisition of new facts can moreover require the updating of the databases considered in the previous point.

- The system needs to deal with different kinds of reasoning.

  There are essentially two complementary ways of reasoning:

  - sequential reasoning, suitable for proposal production; with such a kind of reasoning a sequence of procedures is chosen in a categorical way among the several applicable modules: a procedure is right for a situation or it is rejected;

  - parallel reasoning, suitable for validation of the generated proposals. Often we have to make do with complex alternatives, with a number of advantages and disadvantages which balance one another. The choice is dependent on the weights that are assigned to single facts or relations and implementation of the reasoning can be probabilistic.

In a very schematic way, the support system can be viewed as in Fig. 6.3. The scheme shows a data analysis system (DAS) containing the actual data and all the procedures and methods applicable to them (library of procedures). The external database (DBMS) with its query language (QDBMS) contains information related to the data or to similar analyses. The main body includes an expert system (ES) which must indicate a proposal of procedures to follow, chosen on the basis of information automatically obtained from the actual data in the DAS and/or from the external database DBMS via a language link. The ES is connected to its knowledge base (KB) in which the requirements to activate procedures are mainly specified. The control mechanism performs the planning, that is, generates a list of actions to follow. The output from the ES is the most appropriate sequence of procedures to execute, together with an explanation module about the criteria used in the process of selection of the procedures.

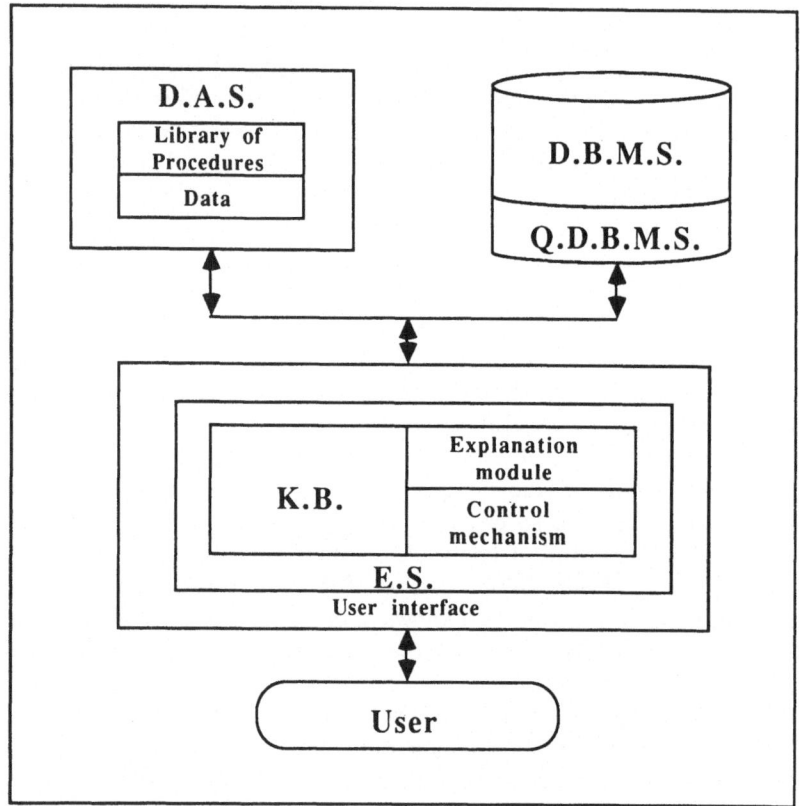

Fig. 6.3. General scheme for a decision support system.

6.4 Which Artificial Intelligence Tools?

In the overall scenario of tools and methods developed to date by experts in artificial intelligence, some seem to be particularly suitable for defining a decision support system.

First of all, there are various techniques for planning, that is, the process of finding a sequence of operations that lead from an initial state to a goal state, under given boundary conditions. The most suitable technique of planning for our case seems to be that of refinement of skeletal plans (Thonnat, 1988; Rosenthal, 1988).

Techniques of machine learning (Tanimoto, 1986) and natural language interfaces (Adorf et al., 1988; Adorf and di Serego Alighieri, 1989) must be taken into account in order to automate the acquisition of knowledge in the knowledge base (KB) and the link to and from the external databases.

Among the various knowledge representations (rules, metarules, semantic networks, frames, and so on) the frame concept seems to be very useful. A frame describes a class of objects or situations on the basis of their different attributes (Minsky, 1975). The advantage in using frames is that attributes of an entity in one frame can be inherited from an entity in another frame. Therefore frames can be very useful in the case of data analysis for taking into account data descriptors, relations, references and procedures related to them.

Knowledge representation via frames can also be combined with the production or rule based approach and with knowledge based systems (Waterman, 1986). In the field of data analysis, the role of an expert system should be to specify, for each analysis, a set of methods which can be applied profitably (Chassery, 1986; Ralambondrainy et al., 1988; Altman, 1985). In it, new methodologies may be added when necessary. The application of the expert system must be integrated with conventional software tools and this capability of integration may condition the success of the entire project. There exist tools oriented towards the production of such a system, expert system tools or EST, which make available a set of tools for use in the building of the knowledge base and also of the particular strategy of inference. In such systems it is possible to have different methods of knowledge representation and to take advantage of both knowledge base and database systems (Bozesan, 1987).

There have not been many applications of artificial intelligence in astronomy to date. Initially, it was mainly classification problems which were examined using artificial intelligence methods. During the last two years, the problem of planning has also been considered in astronomical applications (Albrecht, 1989). Some relevant results have been achieved, as reported elsewhere in this book. These results, systems and/or tools mainly concern planning for data reduction, observational planning with telescopes, choice and control of procedures collected in libraries, intelligent assistance in data analysis systems, language interfaces with external databases, and the link with complex distributed archives.

Although no one of these systems has been built to search for periodicities, it is clear that we can use some of their characteristics for our scope. We plan to study them in more detail to define a system for support of decisions in the general case of the search of periodic signals in astronomy.

References

1. Adorf, H.-M., Albrecht, R., Johnston, M.D. and Rampazzo, R. (1988), "Towards heterogeneous distributed very large databases", *Astronomy from Large Databases: Scientific Objectives and Methodological Approaches*, F.

Murtagh and A. Heck (eds.), European Southern Observatory, Garching bei München, 137–142.

2. Adorf, H.-M. and di Serego Alighieri, S. (1989), "An expert assistant supporting Hubble Space Telescope proposal preparation", *Data Analysis in Astronomy III*, V. Di Gesù, L. Scarsi, P. Crane, J.H. Friedman, S. Levialdi and M.C. Maccarone (eds.), Plenum Press, New York, in press.

3. Albrecht, R. (1989), "Artificial intelligence: what can it do for astronomy?", *Data Analysis in Astronomy III*, V. Di Gesù, L. Scarsi, P. Crane, J.H. Friedman, S. Levialdi and M.C. Maccarone (eds.), Plenum Press, New York, in press.

4. Altman, N. (1985), "Expert systems and statistical expertise. Part I: Statistical expert systems", Department of Statistics, Stanford University, Technical Report 17.

5. Bozesan, M. (1987), "KEEconnection: a technical overview", IntelliCorp Technical Article, IntelliCorp Inc.

6. Buccheri, R. and De Jager, O.C. (1988), "Detection and description of periodicities in sparse data. Suggested solutions to some basic problems", Proceedings of the ASI-NATO conference on *Timing Neutron Stars*, Izmir, Turkey.

7. Chassery, J.M. (1986), "Expert systems for data analysis", *Data Analysis in Astronomy II*, V. Di Gesù, L. Scarsi, P. Crane, J.H. Friedman and S. Levialdi (eds.), Plenum Press, New York, pp. 273–283.

8. De Jager, O.C. (1987), "The analysis and interpretation of VHE gamma ray measurements", PhD thesis, University of Potchefstroom, South Africa.

9. Maccarone, M.C. and Buccheri, R. (1988), "Search for periodicities in gamma-ray astronomy. Towards a decision support system", *Proceedings of the 12th IMACS World Congress on Scientific Computation*, Paris, France, Vol. 2, pp. 322–324.

10. Maccarone, M.C. and Buccheri, R. (1989), "Decision problems in the search for periodicities in gamma-ray astronomy", *Data Analysis in Astronomy III*, V. Di Gesù, L. Scarsi, P. Crane, J.H. Friedman, S. Levialdi and M.C. Maccarone (eds.), Plenum Press, New York, in press.

11. Minsky, M. (1975), "A framework for representing knowledge", *The Psychology of Computer Vision*, P.H. Winston (ed.), McGraw-Hill, New York.

12. Ralambondrainy, H., Demonchaux, E. and Jomier, G. (1988), "Data analysis, databases and expert systems: the common interface", *Astronomy from Large Databases: Scientific Objectives and Methodological Approaches*, F. Murtagh and A. Heck (eds.), European Southern Observatory, Garching bei München, 213–226.

13. Rosenthal, D.A. (1988), "Applying artificial intelligence to astronomical databases — a survey of applicable technology", *Astronomy from Large Databases: Scientific Objectives and Methodological Approaches*, F. Murtagh and A. Heck (eds.), European Southern Observatory, Garching bei München, 245–259.

14. Tanimoto, S. (1986), "Artificial Intelligence — Course Notes", Department of Computer Science, University of Washington, Seattle, WA.

15. Thonnat, M. (1985), "Automatic morphological description of galaxies and classification by an expert system", Rapport de Recherche INRIA no. 387.
16. Thonnat, M. and Clément, V. (1988), "OCAPI: a monitoring tool for the automatic control of image processing procedures", *Proceedings of the 12th IMACS World Congress on Scientific Computation*, Paris, France, Vol. 2, 318–321.
17. Waterman, D.A. (1986), *A Guide to Expert Systems*, Addison-Wesley, New Jersey.

Classification

7 Classification and Knowledge

Michael J. Kurtz
Harvard-Smithsonian Center for Astrophysics
60 Garden Street
Cambridge, MA 02138
USA

(I have tried) "to construct a typology, a model, or perhaps a grammar which will help us to pin down the meaning of certain key words, or of certain evident realities, ..."

Fernand Braudel, *Civilization and Capitalism*

7.1 The Classification Problem

7.1.1 What is Classification?

Classification is the scientific extension of the basic human tool for the expression of knowledge: verbalization. Systems of classification form the vocabularies for the languages of description required by scientific research.

Object classification is the ordered idealization of observational experience; to be successful a classification scheme must also be an ordered idealization of physical reality. Object classification schemes provide the nouns for the formal languages required to describe phenomena.

Automating the classification process, including the creation of classification schemes, is a problem far in excess of current capabilities. Automated procedures to classify objects, however, are quite possible, given a suitably designed classification scheme. The interaction of humans with machines to explore the unknown regimes implied by observations should clearly be the goal of current morphological research; systems which merely label objects with labels from existing classification schemes should be reserved for observations which have inadequate resolution or signal to noise to enable more detailed treatment. We should be attempting to build cognitive systems, not recognitive systems.

7.1.2 Language, Classification, and Perception

As the amount and complexity of the data available to us increases rapidly, the so-called "firehose of data" (McCormick et al., 1987), it is certain that ever increasing responsibility for the basic perception and evaluation of these data will be given over to machines (Kurtz, 1988a). In designing the machine intelligence techniques required by the data it is crucial to recognize that they will impose fundamental restrictions on our ability to know.

Aristotle (1984) suggested that we cannot think without having perceived, and that the scope of our thought is determined by the extent of our perceptual experience. Examples of this from anthropological linguistics are commonplace, such as the Eskimo tribe whose language has 17 different words for different kinds of snow, none of which can be distinguished as anything but snow to outsiders.

Perhaps Kant (1976) best expressed how restrictions in our ability to perceive affect our ability to know. He said that it is not necessary that man's knowledge conform to objects, but that objects must conform to man's apparatus of knowing. It is no coincidence that the discovery of giant arcs caused by gravitational lensing (Lynds and Petrosian, 1987) was made after theoretically generated pictures of them, even in popular films (Schneps et al., 1986), were common.

Object classes may be viewed as ideal Platonic Forms; objects which exist only in the mind, but which are invoked by our perceptual experience, as per the parable of the cave (Plato 1962). This relationship between the worlds of ideas and data forms the basis for the unifying view of the pattern recognition problem, (the lowest level, and most amenable to machine solution, portion of the classification problem), by Simon, Backer, and Sallentin (1982, hereafter SBS) shown in Figure 7.1.

In the *Metaphysica* Aristotle (1984) proposed that the Forms exist in perceptible things, for how could astronomers believe in another "heaven besides the sensible heaven?" This view is carried over into classification as standard objects, "perfect" examples of the classes, used to compare unknown objects against. The existence of standards does not eliminate the existence of the more abstract Form, as the occasional banishment of a standard object, (such as Sirius in the case of the MK spectral classification), because it is not a perfect enough representation of the class, clearly shows. Because of its great beauty and clarity of exposition the MK-78 classification scheme (Morgan et al., 1979), in which single "perfect" objects are used as anchor points in the classification space, represents the closest Aristotelian realization of a system of Platonic Forms yet achieved in science.

A classification system based on standard objects leads directly to the study of representative specimens as a basis for the advancement of theoretical understanding. This is of substantial importance, for as Mihalas (1985) has pointed out "it is specimens, not facts, that are the ultimate empirical currency that we must use if we wish to purchase a valid theory".

While the most important feature of a classification scheme is surely that it span the observational space, the exact details of the ordered systemization of experience are also quite important. For example in the English scheme of object classification a particular class is represented by the word "chair", while in the

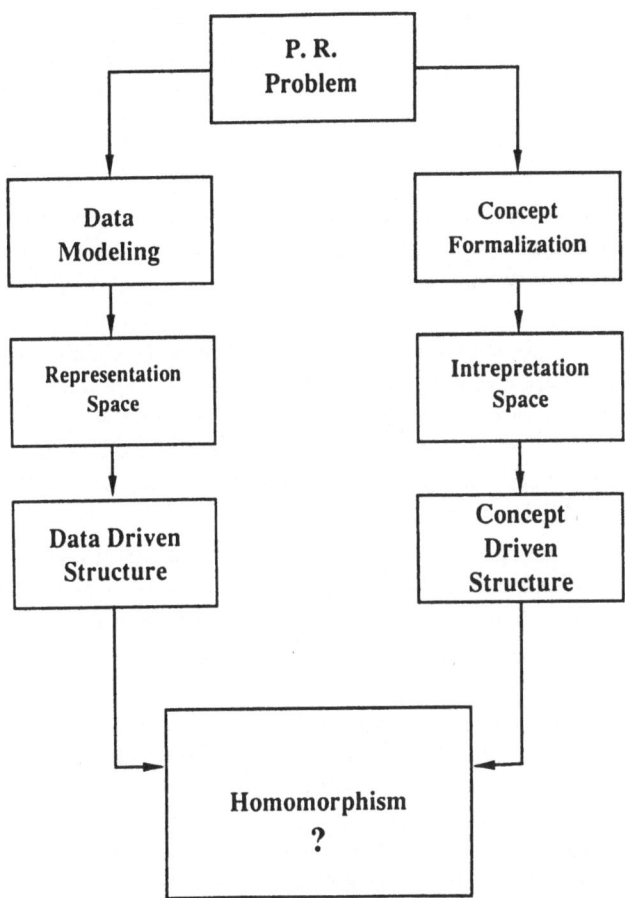

Fig. 7.1. The general solution to the Pattern Recognition Problem, from SBS.

German scheme the word "Stuhl" covers most of the same observation space as "chair", but there is also the word "Sessel" in the German scheme, which is an upholstered chair. The different schemes lead to different types of systematic errors; in English the vagueness of the term leads to the possibility of incorrect assumptions being made, while in German the exactness of the terms leads to a large region of the observational space (that of partially upholstered chairs) where the proper designation is purely a matter of personal opinion.

The relationship between language, classification and perception is not just valid as an abstract ordering of thought and knowledge. Arguably the most important advance in the practical automation of classification came from the field of linguistics: namely Chomsky's (1957) development of generative grammars. This lead directly to hierarchical classification methods on the linguistic model: the syntactic pattern recognition technology of Fu (1974, 1982) and others, which is at the bottom of modern attempts to develop classification schemes using machine intelligence. Accomazzi et al. (1989) discuss a direct astronomical application of these techniques.

7.1.3 Metric Spaces

If, following SBS, we view the classification process as a mapping from the representation (observation) space to the interpretation space, then it is incumbent upon us to inquire into the nature of these spaces, before attempting to create mappings between them.

It is reasonably clear that the observation space is an infinite dimensional vector space, or Hilbert space (Courant and Hilbert, 1968); this is due to the nature of the observations as physical entities. For example the set of all possible spectra are clearly a subset of all functions in $\mathcal{L}_2$ which are of bounded variation. As these functions have Fourier expansions they can trivially be viewed as points in a Hilbert space (Titchmarsh, 1937).

Kurtz (1982) viewed spectra as points in an n-dimensional Euclidean space, and established a classification based on the minimum distance between unknowns and standards. The view of the observation space as a vector space lies at the heart of parametric methods of statistical comparison, such as techniques based on χ^2.

It is also clear that the interpretation space, the space of human ideas and perception, does not have all the pleasant properties of a vector space. While it may be possible to define local measures of perceptual nearness, as Kurtz (1982) has pointed out, these do not in general have the properties of a true metric. In a perceptual space the transitivity of closeness, or the Schwartz inequality, is not necessarily true. For example a dog is not equally (or less) dis-similar to a vacuum cleaner than the sum of the dis-similarities of a dog to a mechanical dog and a mechanical dog to a vacuum cleaner.

The perceptual space does have some nice properties; any perception is very similar to some other possible perception, thus the concept of neighborhood is valid, and different perceptions are distinct, thus the space is separated. Therefore a perceptual space is a Haussdorf space (e.g. Mendelson, 1975).

These considerations lead to the form that machine classification schemes must take. One cannot expect a single similarity measure to be valid in the interpretation space, measures of similarity will be local. This implies that an iterative scheme, with a changing metric, is the proper solution. Kurtz (1982, 1985) has proposed such a scheme for spectral classification; in the parlance of descriptive psychology this is known as a redescription structure (Ossorio 1978, Schideler 1988). The nature of the observation space gives us confidence that this type of iterative solution will converge.

7.1.4 Points of View

Classification is not measurement in the normal sense of that word, rather it is an organizational synthesis of many measurements. As the details of the organization are always a matter of opinion, classification schemes, and individual classifications, involve human judgement. In other words they embody a point of view.

It is normal practice, when dealing with classification data, to take into account the systematic differences between different classifiers, such as the differences between de Vaucouleurs and Lauberts in classifying galaxies. The differences between different spectral classifiers used to be regularly tabulated and published, the so called "personal equation". A claim regularly made for machine solutions to classification problems is that they will eliminate the personal equation. Certainly they will not do that; each machine will have its own "personal" equation. The machine solutions will however eliminate (or at least give control over) the time variation of the personal equation.

Classification schemes have been developed which specifically encode personal judgement. Ossorio (1966) developed a multi-dimensional Euclidean subject matter space for the classification of bibliographic materials. In his scheme a matrix of the relevance of terms as a function of subject matter encodes the human judgement.

The context that an object is found in is important to determining what that object is. Kurtz' (1985) classification scheme gives a direct procedure for determining context dependant numerical values for the perceived importance of a set of measures, given a catalogue of final classifications. LaSala (1988a) is implementing this method to achieve a classification into the MK System of Spectral Classification based on the catalogue of Houk (1978).

Figures 7.2 and 7.3, from LaSala (1988b), show how small differences in the context change Houk's perception of what the important features are to the detailed classification. In Figure 7.2 we see the relative importance of features in the following context: we are sure the spectral type is B9, and think the luminosity class is probably V. In Figure 7.3 we are still sure that the spectral type is B9, but now we think the luminosity class is probably II.

7.2 Epistemology and Classification

While the SBS formulation of the pattern recognition problem in Figure 7.1 is surely correct, it cannot be adequate to describe the classification problem. There cannot be an exact homomorphism between the spaces of data and ideas. This is due to the nature of classification as scientific inquiry. The entire purpose of scientific research, whether morphological or otherwise is to increase our knowledge, to change the idea space.

A system diagram for morphological research must be dynamic, just as scientific research itself is dynamic. Figure 7.4, from Kurtz (1983), shows the SBS formulation modified to show how classification is used to solve physical problems. The feedback paths allow observations to modify our theoretical understanding; they also permit our theoretical beliefs to influence they way we

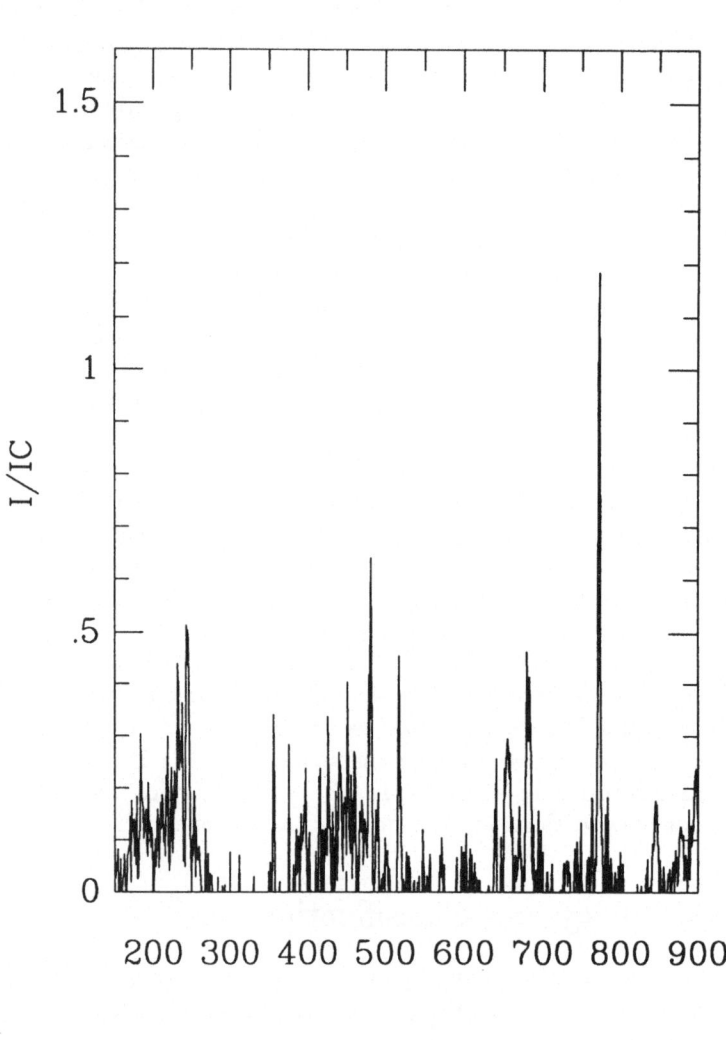

Fig. 7.2. Weighting factors corresponding to classification relevance of spectral features for a spectrum which is surely B9, and probably has a Luminosity Class of V. From LaSala (1988b).

B9I WEIGHT

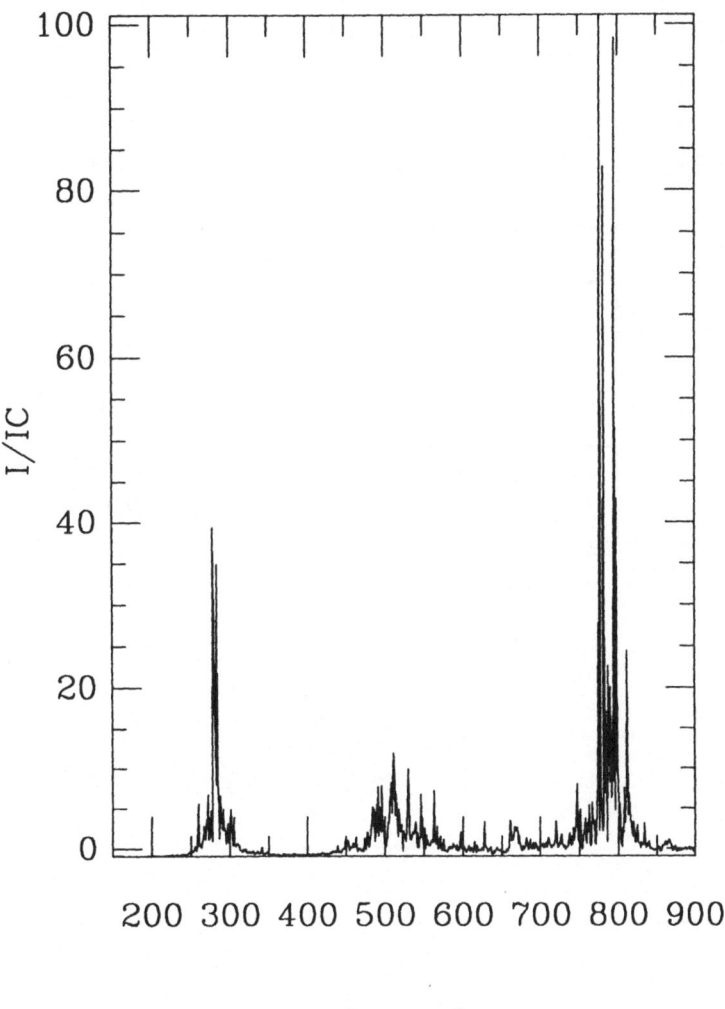

Fig. 7.3. Same as Fig. 7.2, but the Luminosity Class is probably II.

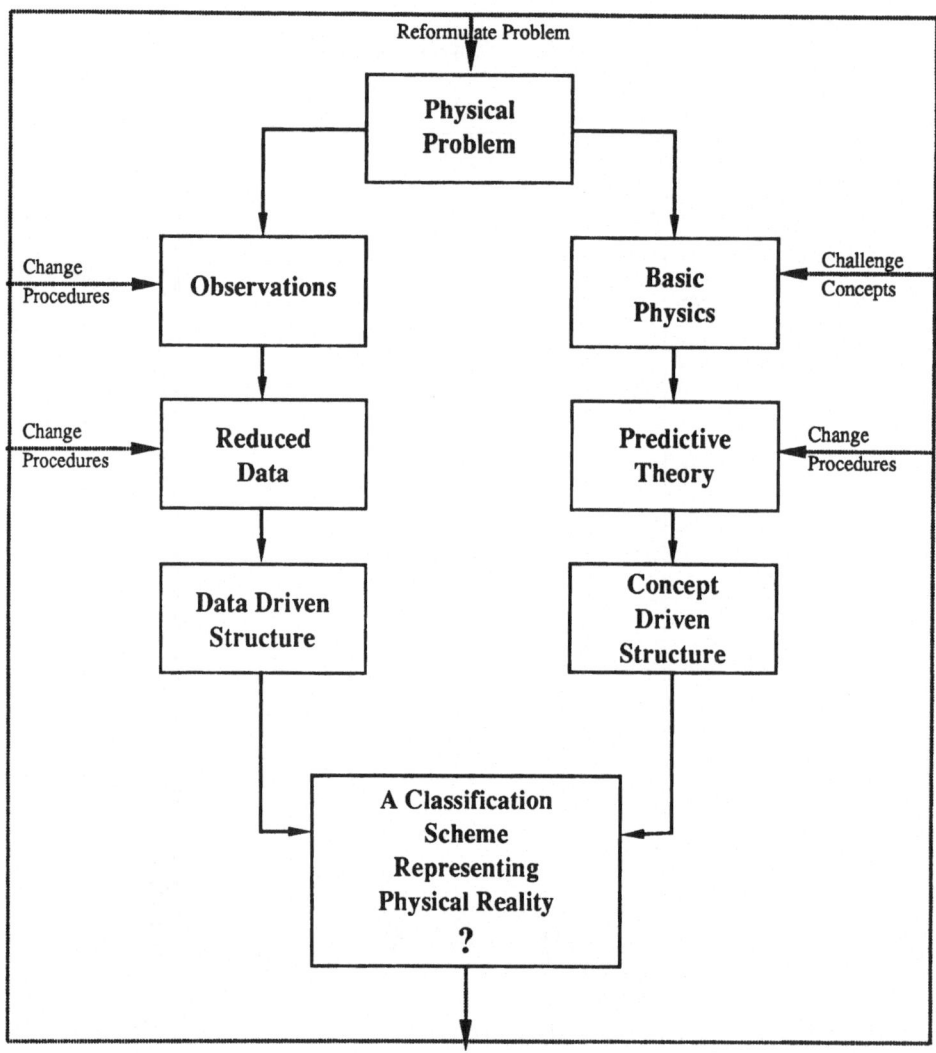

Fig. 7.4. The SBS diagram, modified into a system diagram for morphological research. From Kurtz (1983).

see the data. This is a two edged sword, systematic effects unrelated to the "interesting" physical properties (such as the effect of interstellar material on stellar spectra) can be eliminated, but systematic errors in our understanding can be enhanced.

The system in Figure 7.4 is quite flexible in allowing change, it can respond to incremental advances in our knowledge, but also it can respond to fads. This is not the only way to use data to build theories. One can use a rigid system, firmly grounded in the most certain of our ideas. Because of the rigidness of such a system new knowledge can only be incorporated by breaking it.

Morgan (1985) has long advocated using autonomous classification systems, where the only requirement for the classification scheme is that it be self consistent. Figure 7.5 shows a system diagram for Morgan's scheme; there is no route for the theory to influence the data. The only place where theoretical understanding can influence the classification is in the design of the data driven structure. The task of interpreting the objects, complete with errors, is therefore left exclusively to theory.

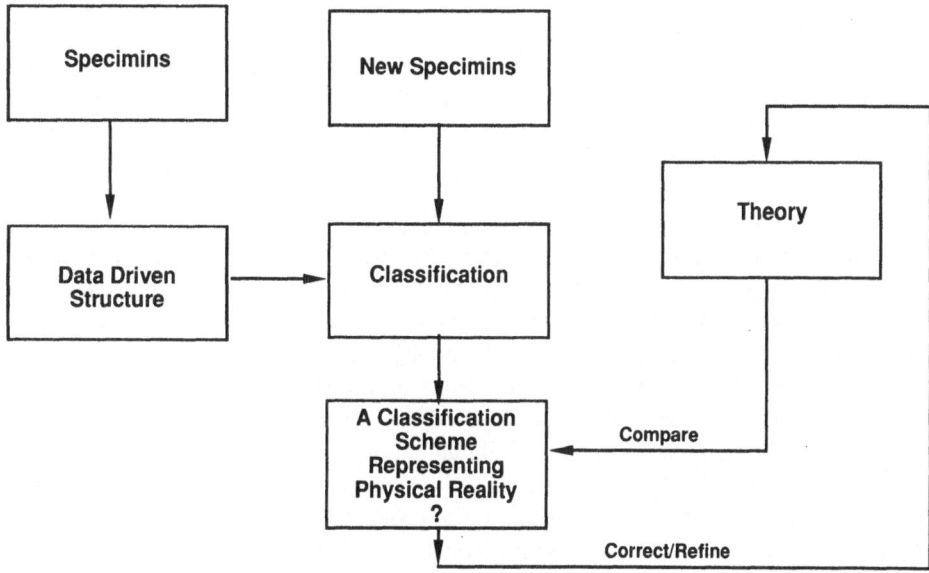

Fig. 7.5. Morgan's scheme for morphological research. From Kurtz (1983).

Both schemes are correct, and in the end both schemes are used. The feedback loop from theory to data is one which is only followed with extreme caution, and the fixed systems must regularly confront other forms of analysis, and when they are found lacking, they are fixed or discarded.

7.3 The Classification of Stellar Spectra

The basic form of the classification scheme for stellar spectra has been well established since the work of Secchi (1867). The modern scheme, shown in Figure 7.6, has been in use since before the beginning of this century, and formed the basis for the Henry Draper catalogue of Cannon (1918–1924). Important lessons concerning the classification process can be gleaned from the early history of spectral classification

Maury (1897), using spectra of higher resolution and signal to noise than Cannon used, created a different classification scheme. Her main types paralleled the OBAFGKM types, but she also noted spectra with certain types of systematic differences with the extra notations "a", "b", "c", and "d". A decade later Herzsprung (1906) noticed that none of the spectra denoted by Maury with "c" (narrow lined) had measured parallaxes, thus they must all be far away and "extraordinarily bright". The concept of giant stars, and of spectroscopic measures of luminosity as well as temperature, thus comes from Maury being careful enough to note a subtle, and totally ununderstood, systematic difference of some of her spectra with "normal" spectra.

The lesson for modern, machine based, classification systems is clear. It is not satisfactory to classify objects into existing classification boxes; one must also provide measures indicating any differences between the properties of the object and the typical properties of the objects in the box. This extra step is of crucial importance if the observations are to lead to increased physical understanding.

Kurtz' (1982, 1985, 1988a, LaSala 1988a, 1988b) scheme provides for the establishment of explicit measures both of the differences between the properties of an individual object and the aggregate properties of the objects in the box to which it is assigned, and of the inability of the classification scheme to account for some of the aggregate properties of the objects in the boxes.

Because it relies on the statistical properties of objects in boxes pre-classified in a catalogue, Kurtz' method uses standard spectra in a manner more similar to Keenan (1987) where many standards define a box, than to Morgan et al. (1979), where a single standard defines a class. If the mean values of the properties of the defining objects in a box are identical to the properties of the principal standard, then the difference in the approaches vanishes. If however this is not the case, perhaps due to the effects of so called "third parameters", then a problem of proper definition and zero point calibration will exist and need to be addressed.

For spectra the technology which requires automation is the combination of the objective prism Schmidt telescope and fast measuring engines. The intrinsic limits in terms or resolution and signal to noise of this technology has tended to drive the classification schemes, which have mostly been rough classifications of low resolution data (for review see Kurtz 1988b). The next generation of

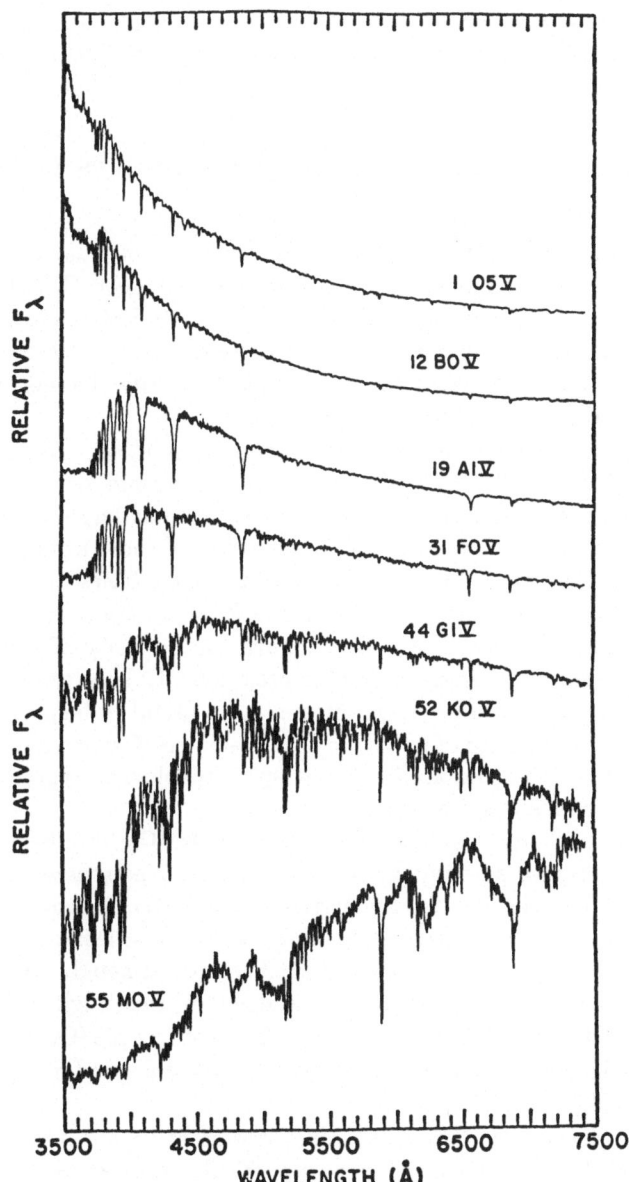

Fig. 7.6. The principal spectral types. From Jacoby et al. (1982).

instruments, spectrographs capable of taking very many simultaneous spectra, is currently under intense development. It can reasonably be expected that survey programs, observing hundreds, perhaps even a thousand, simultaneous high signal to noise, high resolution slit spectra, will be in regular production operation before the beginning of the next century. The millions of spectra coming from these surveys will clearly drive the future classification effort.

7.4 The Classification of Resolved Images of Galaxies

The technology driving the automated classification of galaxies has also been Schmidt telescopes and measuring engines, now with deep CCD frames in very small fields as well. These data are also of relatively low resolution and signal to noise, thus the classification schemes currently in use merely discriminate between stars and galaxies, with no attempt to classify the morphology of the galaxies (see Kurtz 1983, 1988a for reviews).

Technological change in this area is already upon us. New, large format (2048 × 2048 element) CCDs already exist, and should be in regular use on telescopes within one year. These devices will permit large area surveys with depths and signal to noise ratios greatly in excess of those from Schmidt plates. In addition the impending launch of the Hubble Space Telescope, with the Wide Field/Planetary Camera, signals a totally new era in object resolution.

Both of these developments point toward the necessity of developing automated classification schemes of greater precision. Galaxy types have been in the past, and promise to be in the future, quite useful in helping to determine the large scale properties of space (for discussion see Ossorio and Kurtz 1989). The automated classification of galaxies by morphological type however lies in its infancy (for review see Kurtz 1988a).

Kurtz, Mussio, and Ossorio (1989, hereafter KMO) have begun a new program to establish an automatic cognitive system for astronomical image interpretation. A flow diagram for the system, from KMO, is shown in Figure 7.7. The diagram shows the system as an image indexing scheme; the final product of the indexing is not a catalogue of classifications, although classifications can be derived from the catalogue, it is a catalogue of rich, high level descriptors. Using modified versions of the top down judgement module (Ossorio et al., 1988) future investigators should be able to search the catalogue for types of galaxies not yet named.

The KMO scheme is basically a redescription structure (Ossorio and Kurtz 1988b); pixel data is redescribed using statistical and syntactic pattern recognition methods into object primitives, these are redescribed using L-functions (Accomazzi et al., 1989) into structures, which are then evaluated and redescribed into human descriptions, using synthetic judgement techniques from descriptive psychology (Ossorio et al., 1988). A full discussion of the diagram, with its feedback loops can be found in KMO.

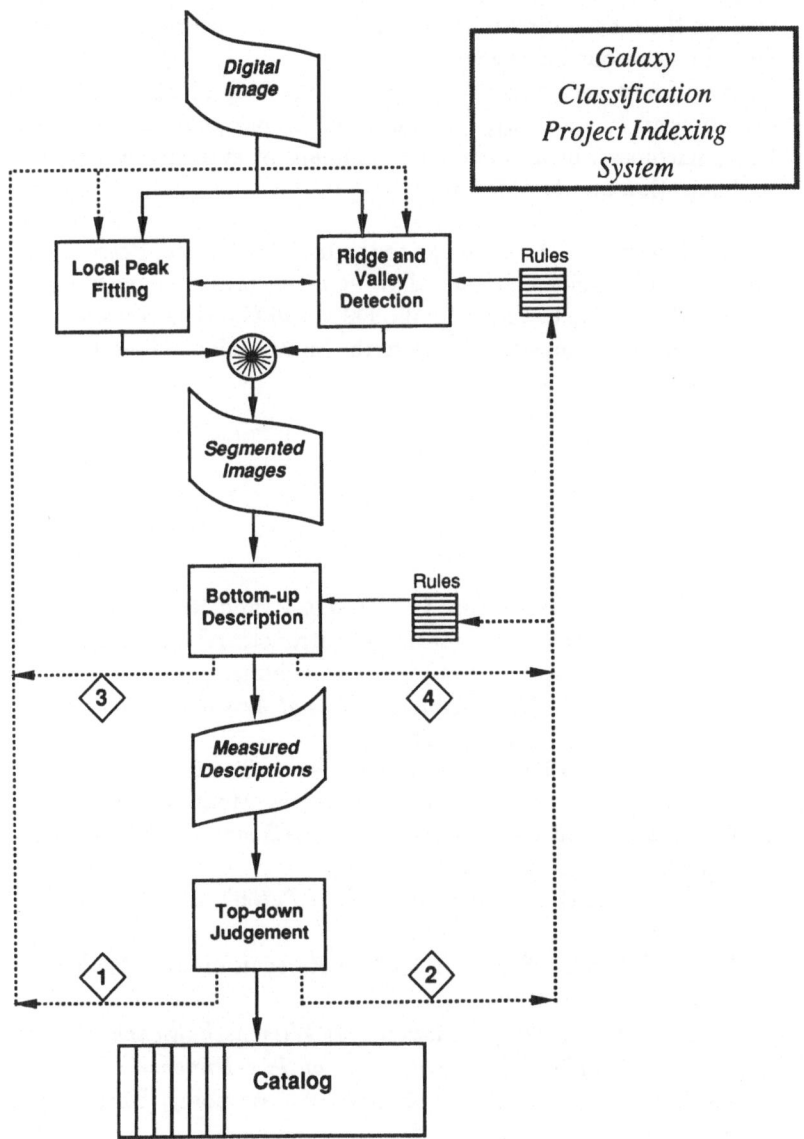

Fig. 7.7. The KMO system for automated morphological galaxy cognition.

7.5 Conclusions

We stand at the edge of a new era; one in which our ability to take data far outstrips our ability to reduce it by human intensive means. We must develop methods of analyzing and classifying our observations so that we do not just arrange them according to what we know, but describe them in ways that allow us to increase our knowledge.

It will be a very long time before machines can look at data with the genius of a Morgan or a Zwicky, but that must be our goal. To achieve this we must build machines which interact with humans at ever increasingly higher levels of sophistication and knowledge.

Acknowledgements. I should like to thank Piero Mussio, Peter Ossorio, and Jerry LaSala for discussions, and for allowing me to present some of their unpublished work. I should like to thank Emilio Falco and Gabriele Germann for discussions. Partial financial support came from the Smithsonian Scholarly Studies Program, and NAGW-201.

References

1. Accomazzi, A., Bordogna, G., Mussio, P. and Rampini, A. (1989), "An approach to heuristic exploitation of astronomers' knowledge in automatic interpretation of optical pictures", this volume.
2. Aristotle (1984), *The Complete Works of Aristotle*, translated by W.D. Ross, Chicago University Press, Chicago.
3. Cannon, A.J. (1918–1924), *Harvard Ann.*, 91–99.
4. Chomsky, N. (1957), *Syntactic Structures*, Mouton, The Hague.
5. Courant, R. and Hilbert, D. (1968), *Methoden der Mathematischen Physik*, 3rd ed., Springer-Verlag, Heidelberg.
6. Fu, K.-S. (1974), *Syntactic Methods of Pattern Recognition*, Academic Press, New York.
7. Fu, K.-S. (1982), *Syntactic Pattern Recognition and Applications*, Prentice-Hall, Englewood Cliffs.
8. Herzsprung, E. (1906), *Zeitschrift für Wissenschaflichen Photometrie*, **3**, 429.
9. Houk, N. (1978), *Michigan Catalog of Two-Dimensional Spectral Types for the HD Stars*, Vol. 2, Department of Astronomy, University of Michigan, Ann Arbor.
10. Jacoby, G.H., Hunter, D.A. and Christian, C.A. (1983), "A Library of Stellar Spectra", Kitt Peak National Observatory, Tucson.
11. Kant, I. (1976), *Kritik der reinen Vernunft*, Suhrkamp, Frankfurt.
12. Keenan, P.C. (1987), "Spectral types and their uses", *Publications of the Astronomical Society of the Pacific*, **99**, 713–723.
13. Kurtz, M.J. (1982), "Automatic Spectral Classification", Thesis, Dartmouth College.

14. Kurtz, M.J. (1983), "Classification methods: an introductory survey", *Statistical Methods in Astronomy*, E. Rolfe (ed.), European Space Agency Special Publication SP-201, 47–58.

15. Kurtz, M.J. (1985), "Progress in automation techniques for MK classification", *The MK Process and Stellar Classification*, R.F. Garrison (ed.), David Dunlap Observatory, Toronto, 136–152.

16. Kurtz, M.J. (1988a), "The search for structure: object classification in large data sets", *Astronomy from Large Databases: Scientific Objectives and Methodological Approaches*, F. Murtagh and A. Heck (eds.), European Southern Observatory, Garching bei München, 113–126.

17. Kurtz, M.J. (1988b), "Report to Commission 45 (Stellar Classification): IV Automated Spectral Classification", *Reports on Astronomy*, **XXA**, J.-P. Swings (ed.), Kluwer, Dordrecht, 637–638.

18. Kurtz, M.J., Mussio, P., and Ossorio, P.G. (1989), "An automatic cognition system for astronomical image interpretation", *Pattern Recognition Letters*, to appear.

19. LaSala, J. (1988a), "A program for automatic two-dimensional spectral classification of objective-prism spectra", *Astronomy from Large Databases: Scientific Objectives and Methodological Approaches*, F. Murtagh and A. Heck (eds.), European Southern Observatory, Garching bei München, 127–133.

20. LaSala, J. (1988b), Invited Paper presented to special session on *Automatic Spectral Classification*, IAU Commission 45, IAU 20th General Assembly. Summary to appear in *Highlights of Astronomy*, R.F. Garrison (ed.).

21. Lynds, R. and Petrosian, V. (1987), "Giant luminous arcs in galaxy clusters", *Bulletin of the American Astronomical Society*, **18**, 1014.

22. Maury, A.C. (1897), *Harvard Ann.*, **28**.

23. McCormick, B.H., DeFanti, T.A., and Brown, M.D., eds. (1987), "Visualization in Scientific Computing", *Computer Graphics*, **21**, No. 6 (special issue), ACM SIGGRAPH, New York.

24. Mendelson, B. (1975), *Introduction to Topology*, Allyn and Bacon, Boston.

25. Mihalas, D. (1985), "On the relevance of the MK system and process to the theory of stellar atmospheres, *The MK Process and Stellar Classification*, R.F. Garrison (ed.), David Dunlap Observatory, Toronto, 4–16.

26. Morgan, W.W. (1985), "The MK system and the MK process" *The MK Process and Stellar Classification*, R.F. Garrison (ed.), David Dunlap Observatory, Toronto, 18–25.

27. Morgan, W.W., Abt, H.A., and Tapscott, J.W. (1979), *Revised MK Spectral Atlas for Stars Earlier than the Sun*, Yerkes and Kitt Peak Observatories, Chicago and Houston.

28. Ossorio et al. (1988), *Knowledge Dictionary System*, Ellery Systems Corporation, Boulder.

29. Ossorio, P.G. (1978), *What Really Happens: The Representation of Real World Phenomena*, University of South Carolina Press.

30. Ossorio, P.G. (1966), "Classification space: a multivariate procedure for automatic document indexing and retrieval", *Multivariate Behavioral Research*, **1**, 479–524.

31. Ossorio, P.G. and Kurtz, M.J. (1989), "Automated Classification of Resolved Galaxies", *Data Analysis in Astronomy III*, V. di Gesù, L. Scarsi, P. Crane, J.H. Friedman, S. Levialdi and M.C. Maccarone (eds.), Plenum Press, New York, in press.

32. Plato (1962), *The Republic*, translated by C.M. Cornford, Oxford University Press, New York.

33. Simon, J.C., Backer, E., and Sallentin, J. (1980), "A Structural Approach of Pattern Recognition", *Signal Processing*, **2**, 5–22.

34. Schneps, M.H., Falco, E.E., Latham, J., and Kurtz, M.J. (1986), Einstein U.S.A. 15", Wolbach Image Processing Laboratory, Cambridge, MA.

35. Secchi, A. (1967), *Memorie della Società Italiana della Scienze*, **1**, p. 47.

36. Shideler, M. (1988), *Persons, Behavior and the World*, University Press of Alabama.

37. Titchmarsh, E.C. (1937), *The Theory of Functions*, Oxford University Press, London.

8 WOLF — A Computer Expert System for Sunspot Classification and Solar Flare Prediction

Richard W. Miller
Cedar Valley Solar Observatory
Hillsburgh, Ontario
Canada, N0B 1Z0

8.1 Introduction

The purpose of my work with WOLF was to provide a means of predicting solar flares using data which an amateur observer could easily obtain. As an amateur radio operator, licence VE3CIE, I was interested in solar flare prediction and the relationship of solar flares to radio communication via ionospheric propagation. The intensity of 1–8 Å X-ray emissions from solar flares can be related to ionospheric propagation phenomena (Reid, 1972).

In 1966, Patrick S. McIntosh of NOAA Space Environment Laboratory, Boulder, Colorado introduced a scheme of sunspot classification (McIntosh, 1986). The McIntosh classification has since been adopted as the standard for the international exchange of solar geophysical data (I.U.W.D.S. Code Book, 1969). The new scheme was a modification to the Zurich sunspot classes which had already been correlated with solar flares (Waldmeir, 1955). The McIntosh classification was designed to incorporate further structural and dynamic aspects of sunspot groups that were noted by observers and flare forecasters to enhance the correlation with flares. McIntosh (1986) and Kildahl (1980) examined the relationship between sunspots and flare X-ray emission. Their results indicated that the McIntosh classification makes a meaningful distinction between flare-active and non-flaring active regions.

Since observations of sunspots in white light are easily made by amateur astronomers, the McIntosh classification scheme promised to provide a means of predicting the occurrence of solar flares that might affect radio communications. The requirement was to be able to recognize the sunspot features and allocate the sunspots to one of the sixty-three McIntosh classes. The class could then be related to the potential for X-ray flare production as determined from statistical analysis of previous activity of similar spot groups. To this end, I prepared "A Guide to Sunspot Classification and Solar Flare Prediction" (Miller, 1987) which contained a key to allocating sunspot groups to McIntosh classes and the results of an analysis of 12,411 sunspot groups and 1485 associated flares. McIntosh (1987) pointed out that the Guide's format for associating sunspot characteristics with flare probabilities came close to the arrangement of decision trees in an expert system built in a joint NOAA/University of Colorado project (Lewis and Dennett, 1986). The expert system was named THEO, after Theophrastus, a disciple of Aristotle, who may have been the first to record having seen a sunspot in 325 B.C..

THEO predicts X-ray flares for the next 24 hours based on a combination of McIntosh sunspot classification and a number of dynamic qualities of sunspot groups which would be difficult to incorporate into the classification. This information prompted me to attempt to develop a small expert system for amateur use.

Expert systems are non-traditional computer programs that consist of three major parts:

(1) a data base of knowledge, called the knowledge base, about the application domain, which is derived from the knowledge of one or more human experts. This knowledge is often encoded in the form of IF-THEN rules.

(2) a working memory which consists of current facts about the application domain (e.g. the current maturity and shape of the penumbra on the largest spot in an active region under observation). This information is usually supplied by the user who is prompted by the program when the data are required. In sophisticated systems, the information may be supplied by external sensors.

(3) the inference engine which is the computational part of the program. The inference engine applies the rules in the knowledge base to the facts in the working memory in order to draw conclusions. Inference engines may employ one of two search techniques, either forward or backward chaining. Forward chaining starts with a condition and works its way forward through a chain of rules to arrive at a conclusion. Backward chaining starts with a conclusion and works its way backward to find if a set of conditions exist which verify the conclusion. For example: *Foreward Chaining:* If the sunspot group is unipolar with no penumbra then its class is Axx (condition → conclusion). *Backward Chaining:* If the sunspot group is Axx then it is unipolar with no penumbra (conclusion → condition).

8.2 The Knowledge

The knowledge for WOLF was compiled during the research carried out in preparation for writing the Guide. The heuristics (rules of thumb) for classifying sunspots were formalized in the Guide in the key to allocate sunspot groups to McIntosh classes.

Sunspot Group Type = Class + Largest Spot + Interior

D   a   i

where ;   D - modified Zurich class

a - type of largest spot

i - distribution of spots between leader and follower spots

Fig. 8.1. Components of the McIntosh Sunspot Classification Scheme.

8.2.1 The McIntosh Classification of Sunspots

The basis of the McIntosh system (see Appendix Figure 8.A1) is the Zurich classification which attempts to describe a typical evolutionary sequence of large sunspot groups. By the addition of two components, the nine Zurich classes have been expanded to sixty-three McIntosh classes which produce a more detailed description of sunspots. The new classes can be easily determined with little additional effort over the Zurich classes but greatly improve flare/no flare segregation. In order to facilitate the analysis of sunspot groups, a key was devised for the McIntosh classification. The key includes the following characteristics which are required for classification:

1) group polarity
2) group length
3) penumbra on the largest spot

4) maturity and shape of the penumbra
5) penumbra size on the largest spot
6) spot distribution within the group

These features are summarized in the McIntosh Sunspot Classification in three components in Figure 8.1 (McIntosh, 1986). An example of how a decision tree in the key (Miller, 1987) works is given in figure 2.

For the spot shown follow the path as indicated below. The selected spot has been analyzed to be McIntosh classification Hax.

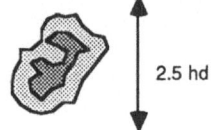

2.5 hd

McIntosh Sunspot Group Classification – Key

Chart 1

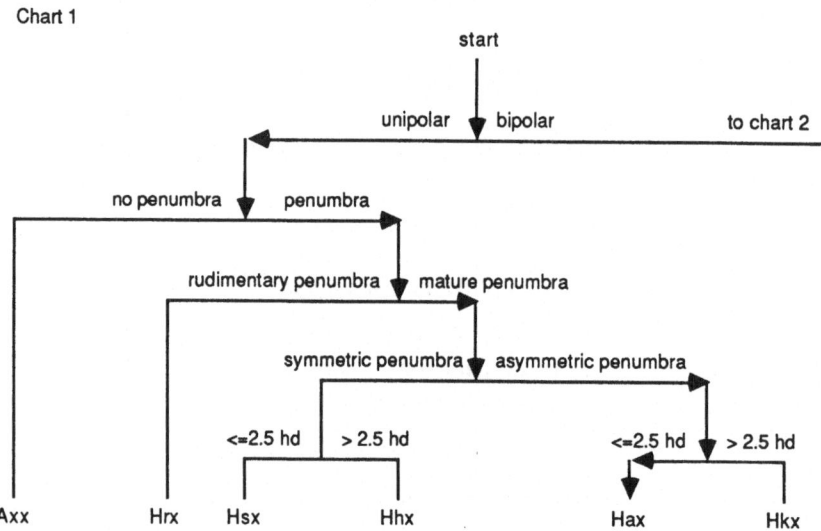

Fig. 8.2. Analysis of a sunspot group using a decision tree from the McIntosh Sunspot Group Classification – Key.

8.2.2 Solar X-ray Flare/Sunspot Relationships

For many years, solar flares were classed according to optical importance (Table 8.1). The optical importance scale is based on the area of the enhanced plage (square heliographic degrees) and brightness measured in Hα (6563 Å).

Table 8.1. Optical and X-ray classification of solar flares.

Optical		SESC X-ray	
Area Importance	Area (square degrees)	X-Ray Class	Peak Energy Flux (Wm^{-2})
S	less than 2.1	A	less than 10–7
1	2.1 to 5.1	B	10–7 to 10–6
2	5.2 to 12.4	C	10–6 to 10–5
3	12.5 to 24.7	M	10–5 to 10–4
4	24.8 and greater	X	$\geq$ 10–4

A brightness qualifier is usually appended to the optical classes: F – faint, N – normal, B – bright.

In 1969, the Space Environment Services Center (SESC), NOAA introduced a classification scheme which ranks solar flares according to their peak X-ray emission in the 1–8 Å wavelength range. Radiation at these wavelengths is associated with ionospheric disturbances which disrupt radio communication. The scheme was intended to supplement, but not replace, the optical classification.

The X-ray classification has two advantages over the optical classification (SESC, 1986):

(1) it gives a better measure of the geophysical significance of the event and
(2) it provides an objective means of classifying almost all geophysically significant activity regardless of its location on the solar disk at or beyond the limb.

A number between 1.0 and 9.0 is appended to the letter as a multiplier. For example, a C3.2 burst indicates a peak X-ray flux of 3.2×10^{-6} Wm^{-2}, whereas X5 would indicate a peak flux of 5×10^{-4} Wm^{-2}. Since the emission is currently measured as the X-ray flux from the entire disk, bursts from an individual region will not be discerned if they do not exceed the background emission from the whole disk. During brief intervals at solar maximum, the background can rise as high as class M and during these times classes A, B, and C cannot be detected. The background reaches class A during solar minimum. The $H\alpha$ optical classes 1, 2, 3 correspond approximately (Table 8.1) to the X-ray classes C, M, X respectively (Hirman et al., 1980).

Using a large database from the descending portion of solar cycle 20 (1969-1976), K. Kildahl of ERL, NOAA examined the correlation between McIntosh sunspot classes and the M and X X-ray flares (Kildahl, 1980). The M and X class flares were catalogued using the 63 McIntosh classes and the number of days each sunspot type appeared on the disk. There were 12,411 classifications of sunspots observed and 1344 class M flares and 141 class X flares.

These data were used in the Guide to compile a set of flare indices representative of the probability of production of M and X class flares for each of the McIntosh classes.

$$\text{Flare Index} = \frac{\text{number of flares observed in the class}}{\text{number of days the class was observed}} \times 100$$

A value of 100 for the index indicates that each time the class occured, an X-ray flare of the type indicated was observed.

Figure 8.3, from McIntosh (1986), based on Kildahl's analysis illustrates the effectiveness of the McIntosh system in discriminating flaring from non-flaring groups in modified Zurich class F. Note that groups Fsi, Fki and Fkc are expected to produce M class x-ray bursts within the next 24 hours, with 100% certainty. According to McIntosh, "No other solar parameter has proven to be as accurate a flare predictor" (McIntosh, 1986). For a detailed discussion of X-ray flare/sunspot relationships, the reader is referred to *Solar-Terrestrial Predictions Proceedings*: R. F. Donnelly (ed.) 1980, Boulder/NOAA-ERL and *Solar-Terrestrial Predictions: Proceedings of a Workshop at Meudon, France*, 1984, P. A. Simon, G. Heckman and M. A. Shea (eds.), 1986, NOAA (Boulder)/AFGL (Bedford).

McIntosh Classification

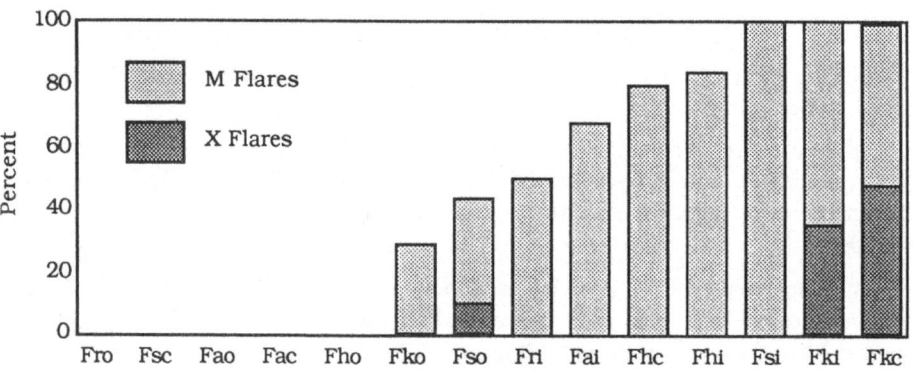

Fig. 8.3. Class F sunspot groups and their correlations with X-ray flares of intensity greater than M class occurring in the next 24 hours (after McIntosh, 1986).

8.3 Development of WOLF

The expert system was named WOLF after Rudolf Wolf (1816–1893), who was director of the Zurich Observatory and who introduced the relative sunspot number R, in 1848 , as a measure of solar activity. WOLF was developed using the expert system shell EXSYS designed by EXSYS Inc. and runs on an IBM PC, AT or compatible. The knowledge of the McIntosh classification scheme was organized in a decision tree (Appendix Figure 8.A2). The decision tree was used to create a set of IF-THEN rules which form the knowledge base for WOLF. In WOLF, knowledge concerning the characteristics of sunspot groups is stored in CONDITIONS.

A CONDITION consists of two parts: a QUALIFIER and one or more associated VALUES. Each CONDITION refers to a specific characteristic of a sunspot group such as polarity or length. For example the condition associated with the sunspot group characteristic length is made up as follows:

QUALIFIER: the length of the sunspot group in heliographic degrees is:

VALUES: 3 to 9 degrees inclusive
10 to 15 degrees inclusive
greater than 15 degrees

The CONDITION for a spot group of modified Zurich class D for the characteristic length would be: The length of the sunspot group in heliographic degrees is 3 to 9 degrees inclusive. WOLF uses six QUALIFIERS and twenty-two VALUES to characterize its knowledge of sunspot groups.

WOLF must determine the solution to three problems: 1) To what McIntosh class does the sunspot group belong? 2) What is the probability of the group producing an M class X-ray flare? 3) What is the probability of the group producing an X class X-ray flare? The possible solutions to the problems are resident in the knowledge base in the form of CHOICES. In WOLF there are sixty-three possible McIntosh classes to chose from and nineteen possibilities for probable flare production.

The choices for M or X type flare production have been related to the Guide flare index as follows (Harmon and King, 1985):

Flare Index Probability of Flare Production

Flare Index	Probability of Flare Production
0	definitely not
1–14	almost certainly not
15–34	probably not
35–64	equal chances
65–79	slight evidence
80–89	probably
90–99	almost certainly
100	definitely

Also, it is possible that WOLF has no knowledge of a particular McIntosh class producing a flare or that the observer has made an error in describing the group to WOLF. The knowledge that allows WOLF to arrive at a solution is represented in a set of sixty-three IF-THEN rules in the knowledge base. Each rule consists of four parts, an IF part, a THEN part, a note and a reference. The IF part consists of a series of conditions which WOLF tests against a description of the sunspot group provided by the observer to see if the conditions are true. The THEN part consists of a series of choices and their associated probability values. The note provides additional information pertaining to the rule such as the number of groups sampled in the statistical study. The reference cites the source of the knowledge contained in the rule.

For example:

The rule for sunspot group Axx is:

IF: The polarity of the sunspot group is unipolar (a single spot or group of spots with the greatest separation between spots < 3 degrees) and the largest spot in the group has no penumbra.

THEN: The McIntosh class of the group under consideration is Axx — probability = 100/100

and this group will almost certainly not produce an M flare — probability = 1/100

and this group will almost certainly not produce an X flare — probability = 1/100

The EXSYS shell has an inference engine which employs a backward chaining technique to search through the rules in the knowledge base to arrive at a solution. When WOLF is run, the program determines which rules are applicable to a specific CHOICE and tests the CONDITIONS in the rules IF part to see if they are true. When the IF conditions in a rule are determined to be true, WOLF assumes that the CHOICES in the THEN part are also true. That is, WOLF has arrived at a solution.

8.4 A Dialogue With WOLF

Perhaps the best way to grasp the operation of WOLF is by looking at a transcript of a dialogue between WOLF and an observer. The dialogue is shown below, with the observer's response in block type. Comments are interspersed throughout.

WOLF begins by providing a brief description of itself.

This expert system (WOLF) is designed to assist you in catagorizing sunspot groups into one of sixty-three possible catagories in the McIntosh sunspot classification scheme. WOLF will then indicate the probable X-ray flare production of the group based on knowledge of past performance of over 12,000 sunspot groups which produced some 1300 flares over a 5 year period in the declining phase of cycle 20.

WOLF will ask a number of questions concerning the following characteristics which are required for classification:

1) group polarity   4) penumbra shape
2) group length   5) size of largest spot
3) existence of a penumbra 6) spot distribution

WOLF will indicate the McIntosh class of the group under consideration and the probability of the group producing both M and X class X-ray flares. The knowledge used by WOLF to determine the flare probabilities was obtained from a study of data by K. Kildahl (1980).

WOLF then requests information concerning the sunspot group the observer is considering. The polarity of the group is an essential piece of information which WOLF requests first.

The polarity of the sunspot group is:

1. unipolar (a single spot or group of spots with the greatest separation between spots < 3 degrees)
2. bipolar (two or more spots forming an elongated cluster of length > 3 degrees. Usually there will be a space near the middle of the cluster dividing it into two distinct parts of opposite magnetic polarity).

The observer replies: **1**
The second piece of information WOLF wants concerns the penumbra on the largest spot.

The largest spot in the group has

1. no penumbra
2. a rudimentary penumbra which partially surrounds the spot
3. a small symmetric penumbra with north to south diameter ≤ 2.5 degrees
4. a small asymmetric penumbra with north to south diameter ≤ 2.5 degrees
5. a large symmetric penumbra with north to south diameter > 2.5 degrees
6. a large asymmetric penumbra with north to south diameter > 2.5 degrees

The observer wants to know why WOLF needs this particular information, so asks: **Why?**
WOLF explains by displaying the heuristic rule it is considering, RULE 1.

Rule Number: 1
IF: 1. The polarity of the sunspot group is unipolar (a single spot or group of spots with the greatest separation between spots < 3 degrees)
and 2. The largest spot in the group has no penumbra.
THEN: The McIntosh class of the group under consideration is Axx —
    probability = 100/100
and This group will almost certainly not produce an M flare —
    probability = 1/100
and This group will almost certainly not produce an X flare —
    probability = 1/100
NOTE: there were 2517 Axx groups in the sample.

The observer then asks WOLF how it knows that a condition in the IF part is true by entering the line number of the IF condition: **1**
WOLF replies: You told me.
The observer asks about condition 2: **2**
WOLF replies: I do not know.
WOLF has indicated that in order to determine if Rule 1 is true it must have information concerning CONDITION 2 in the IF statement (i.e. information about the penumbra on the largest spot).

The observer then returns to the question concerning the penumbra on the largest spot and provides an answer: **4**

At this point WOLF has determined that it has sufficient information to classify the particular group being considered and provides the following answer.

Solar flares are ranked according to their peak X-ray emission in the 1–8 Å wavelength range. This band has been chosen because radiation at these wavelengths is associated with ionospheric disturbances which disrupt radio communication. The Space Environment Services Center receives data from the GOES satellite in the 1–8 Å range 24 hours a day. Class M flares have a 1–8 Å peak energy flux 10^{-5} to 10^{-4} watts-m^{-2} and class X flares $> 10^{-4}$.

Class M and X X-ray flares correspond for the most part to Hα optical classes 2 and 3 respectively. (The number in brackets following the statement of flare probability is the number of flares which occurred per 100 appearances of that particular McIntosh group.)

The McIntosh class of the group under consideration is Hax.
This group will almost certainly not produce an M flare. (6/100)
This group will almost certainly not produce an X flare. (1/100)

8.5 Summary

The McIntosh Classification provides a more meaningful distinction between flaring and non-flaring active regions than the Zurich classification. However, it is an empirical approach based on statistical correlations rather than a thorough understanding of the physics of flare production.

Analysis of nineteen parameters, useful for flare prediction (Neidig et al., 1981) identified the McIntosh classification components as important predictors for flare forecasting. The most useful predictor, however, was persistence (i.e., if there are flares today, there will be flares tomorrow). Even with the inclusion of predictors such as the occurrence of bright points along active region neutral lines, magnetic class, magnetic gradient, intrusion of new magnetic flux into an old active region, and motions between adjacent concentrations of magnetic flux of opposite polarity, to name a few; success in forecasting flare occurrence, 24 hours in advance, is slightly more than 50 per cent.

While it may be possible to improve this score by considering other empirical forecasting rules, it is not likely that significant progress can be made without a substantial improvement in the understanding of flare physics. This will require simultaneous soft X-ray, hard X-ray, microwave, and optical observations at high spatial (≤ 1") and temporal (about 0.1 s) resolution because the important physics is occurring very fast on a very small scale. This capability will not be available to professional flare forecasters for a long time in the future.

Because of the empirical nature of knowledge concerning flare/sunspot relationships, solar flare prediction is a decision making task that depends to a large degree on judgement and experience. Joseph Hirman, chief forecaster at SESC explains, "The present techniques for producing daily forecasts for solar flare activity use actual solar observations but also depend to a significant degree on human judgement ... Forecasting is not an exercise in physics and applied math, it is an exercise in recognizing, scaling, catagorizing and decision making". (Lewis and Dennett, 1986).

Expert systems are useful tools for capturing this type of expertise which relies heavily on symbolic, non-algorithmic methods of problem solving. They exhibit benefits in the following areas:

1. improved data analysis and decision making support through enhanced consistency and thoroughness,
2. provision of a systematic means to formalize and preserve heuristic forms of knowledge and statistical information from case studies,
3. provision of access to expertise for many users

WOLF is an expert system which can classify sunspot groups according to the McIntosh sunspot classification scheme and indicate the probability of flare occurrence. The feasibility of capturing the experience and knowledge of an expert in a specific astronomical domain using expert systems has been demonstrated. At the present time, WOLF is limited to the use of the McIntosh classification as the knowledge base for flare prediction. Its prediction capabilities could be enhanced by incorporating other dynamic qualities of sunspot groups such as was done with THEO.

Acknowledgements. I wish to thank Dr. P. S. McIntosh, NOAA Space Environment Laboratory, Boulder, Colorado, for identifying the similarity between decision trees in an expert system and the format of the key in the Guide and for forwarding to me the Lewis and Dennett paper on the expert system THEO.

References

1. Harmon, P. and King, D. (1985), *Expert Systems, Artificial Intelligence in Business*, Wiley, New York, 42.
2. Hirman, J. W., Neidig, D. F., Seagraves, P. H., Flowers, W. E. and Wiborg, P. H. (1980), in *Solar-Terrestrial Predictions Proceedings*, Vol. 3, R. Donnelly (ed.), US Department of Commerce, C-64.
3. International Ursigram and World Day Service (I. U. W. D. S.) (1969 and subsequent years), *Synoptic Codes for Solar and Geophysical Data*, 2nd Revised Edition, NOAA Space Environment Services Center and Observatoire de Paris, 41.
4. Kildahl, K. (1980), in *Solar-Terrestrial Predictions Proceedings*, vol. 3, R. Donnelly (ed.), US Department of Commerce, C-166.
5. Lewis, C. and Dennett, J. (1986), *CU Engineering*, **3**, University of Colorado, College of Engineering and Applied Science, 15.
6. McIntosh, P. S. (1986), in *Solar-Terrestrial Predictions: Proceedings of a Workshop at Meudon*, France, 1984, P. A. Simon, G. Heckman and M. A. Shea (eds.), US Department of Commerce, 357.
7. McIntosh, P. S. (1987), letter to author, 17 June 1987.
8. Miller, R. W. (1987), "A Guide to Sunspot Classification and Solar Flare Prediction", Cedar Valley Solar Observatory, Hillsburgh, Canada.

9. Neidig D. F., Wiborg, P. H., Seagraves, P. H., Hirman, J. W. and Flowers, W. E. (1981), "An Objective Method for Forecasting Solar Flares", AFGL-TR-81-0026, Air Force Geophysics Laboratory, Hanscom AFB, Mass.

10. Reid G. C. (1972), in *Solar Activity Observations and Predictions*, P. S. McIntosh and M. Dryer (eds.), MIT Press, Cambridge, Mass. 293–312.

11. SESC (1986), Descriptive Text, Contents of *Preliminary Report and Forecast of Solar Geophysical Activity*, US Department of Commerce, 2.

12. Waldmeir, M. (1955), *Ergebnisse und Probleme der Sonnenforschung*, 2nd ed., Geest und Portig, Leipzig.

Appendix
McIntosh Sunspot Classification
Figure 1
hd = heliographic degree
Modified Zurich Class
A
unipolar group, no penumbra early or final stage of evolution
B
bipolar group, without penumbra on any spots
C
bipolar group with penumbra on one end of group, usually the largest of leader umbrae
D
bipolar group, penumbra on spots at both ends, length < 10 hd
E
bipolar group, penumbra on spots at both ends, length10-15 hd inclusive
F
bipolar group, penumbra on spots at both ends, length > 15 hd
H
unipolar group with penumbra usually evolved from a larger group
Type of Largest Spot
x
no penumbra
r
rudimentary penumbra partially surrounding largest spot
s
small, symmetric, north to south diameter <= 2.5 hd
a
small, asymmetric, north to south diameter <= 2.5 hd
h
large, symmetric, north to south diameter >= 2.5 hd
k
large asymmetric, north to south diameter >= 2.5 hd
Sunspot Distribution
x
undefined for unipolar groups
o
open, few spots between leader and follower, weak magnetic field gradient
i
intermediate, many spots between leader and follower, none with penumbra
c
compact, many strong spots between leader and follower, at least one has penumbra

Fig. 8.A1. Appendix Figure 1.

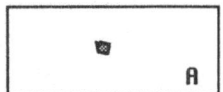

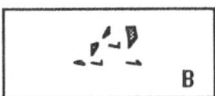

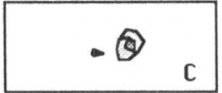

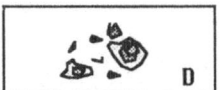

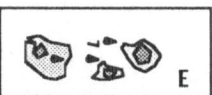

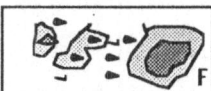

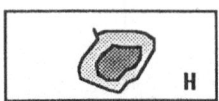

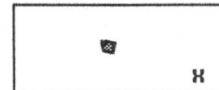

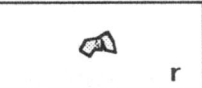

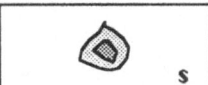

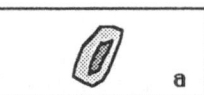

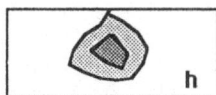

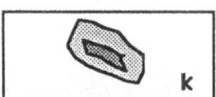

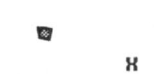

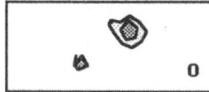

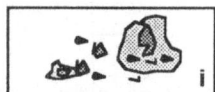

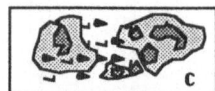

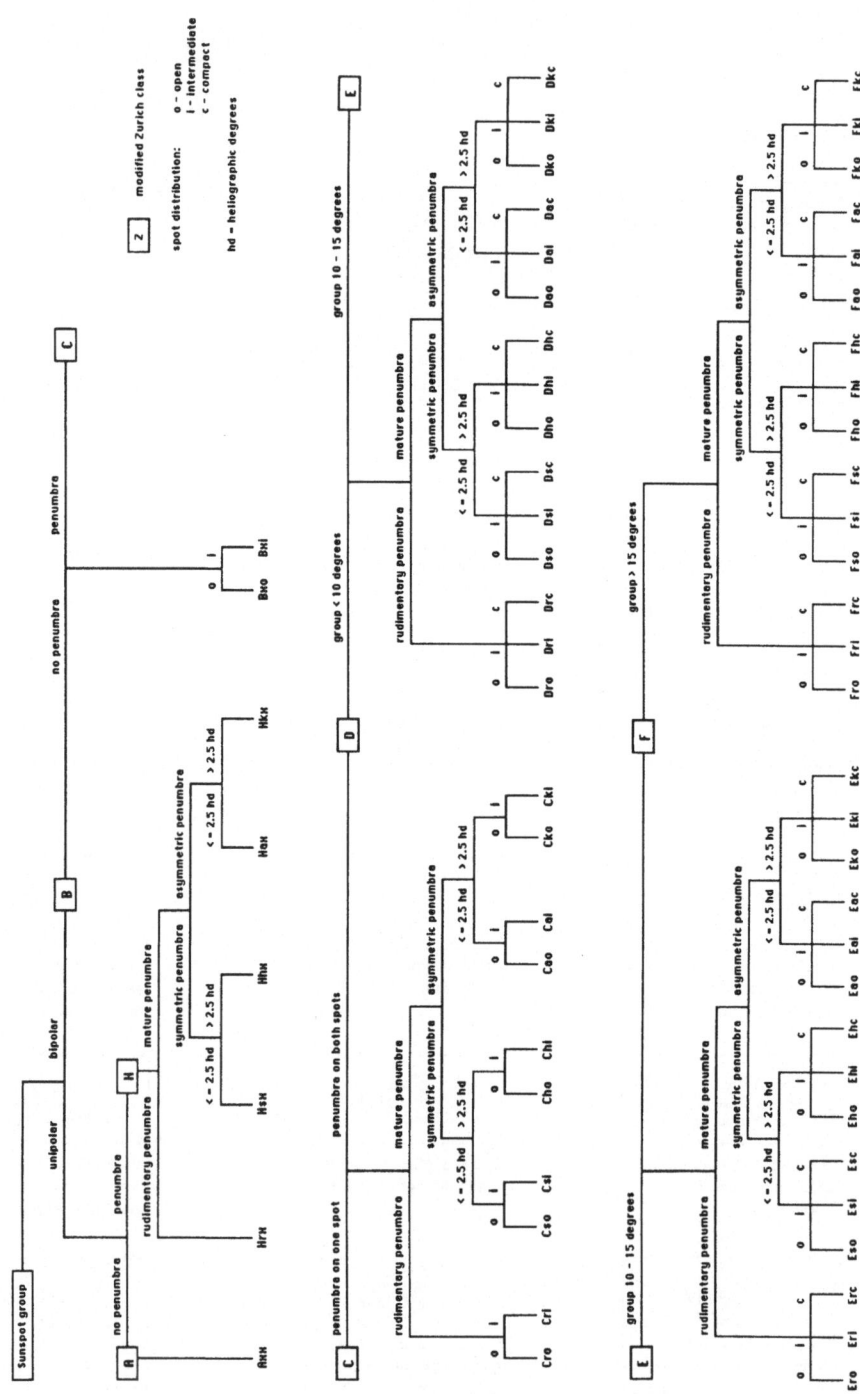

Fig. 8.A2. Appendix Figure 2.

9 Knowledge Based Classification of Galaxies

Monique Thonnat (1) and Albert Bijaoui (2)

(1) INRIA Sophia Antipolis
2004, route des Lucioles
F-06565 Valbonne
France

(2) OCA Observatoire de Nice
Le Mont Gros
Boîte Postale 139
F-06003 Nice Cedex
France

9.1 Introduction

9.1.1 Classification as a Fundamental Tool in Observational Science

The classification of objects belonging to a given category is the essential link between observational data and physical hypotheses. Concepts (classes) allow us to explain these data and to predict which are connected with others. Success of classification lies in the reduction principle, implying that less information is required to determine observables.

A category is always a fuzzy concept. Generally its limits result from some previous classifications. A given classification proceeds from the definition of a given category of objects in a preceding step. In other words, classification is always hierarchic, from the general concept of element up to all kinds of category of similar objects.

Classifications must be open to new kinds of object, or new observational data, giving a new perspective on the categories. The classification building process takes into account the observational data and provides a set of classes.

Each object must belong to one or perhaps more of these classes. Membership can be absolute or fuzzy. In the latter case, we obtain a set of membership probabilities for each object to be classified.

If the classification is carried out well, elements belonging to the same class can be considered as similar objects, and they can be named in a similar way. Classification is also the way our language is enriched, by the introduction of new words associated with new classes of objects. We must not forget that if language results from human thinking, human thinking results also from language. In this sense, object classification is the fundamental tool of observational science.

9.1.2 Generalities on the Classification of Galaxies

The classification of galaxies began before the real identification of galaxies as a major category of astronomical object (Wolf, 1908; Sandage, 1975). Galaxies were a subset of nebulous objects on photographic plates.

Hubble's classification (Hubble, 1926) appeared with the use of the large Mount Wilson telescope. The success of this scheme lies in the small number of bins allowing the classification of the great majority of galaxies, even if peculiar features may mask the structure.

This scheme, and its further improvements (Sandage, 1961; de Vaucouleurs, 1959), allowed one to define the language associated with galaxies and to guide astronomers in their understanding of this category of celestial object.

With the increasing amount of new observations, new classes of objects were identified: galaxies with an active nucleus, galaxies with a jet, radiogalaxies, etc. Nonetheless the revised Hubble scheme is still considered as the first step for putting an observed galaxy in a bin. Other properties are thus considered as a consequence of an abnormal event, more than a defect in the scheme itself.

The classification of galaxies is defined by morphology. Four large classes can be identified: elliptical, lenticular, spiral and irregular galaxies. These classes are subdivided by taking into account other morphological features: ellipticity, bar or ring. Expertise is necessary to give an accurate type to a given galaxy. This work has been carried out for the brightest galaxies (de Vaucouleurs et al., 1976).

9.1.3 The Need for Automated Classification

Current classifications are subjective. The same astronomer can attribute different types to the same object, according to the observational conditions. Only a classification based on well-defined rules applied with a computer can give objective information.

Visual analysis of the image can take a long time, and the number of classified objects cannot increase rapidly. As opposed to this, automated analysis is faster, allowing us to process a great number of objects. Visually classifying galaxies also necessarily means examining fainter objects. Visual analysis fails generally in the transfer of classification rules, adapted to bright objects, to the fainter ones. A well-built automated system, which takes quantitative information into account, must give more relevant information.

Morphological classification of galaxies does not take into account all information known about these objects; instrumental improvements and the use of a set of large telescopes in many wavelength ranges (X-ray, ultraviolet, visible, infrared, radio) allow us to obtain more and more new kinds of data. The use of multivariate analysis by computer permits the building of new statistical classifications based on these data. New classes and new rules can be introduced, leading to another vision of the category of galaxies.

Rendering Hubble's classification more sophisticated by the use of new data is only possible using an approach starting from the scheme to be outlined, and by building new rules allowing classification to be improved. If statistical analysis can extract these rules, only Artificial Intelligence methods allow us to activate these rules, in a knowledge based system.

9.2 Artificial Intelligence Tools for Galaxy Classification

9.2.1 Expert Systems

In this section we present the artificial intelligence tools which can enable the building of an automated galaxy classification system: expert systems (Hayes-Roth et al., 1983; Hanson and Riseman, 1978; Shortliffe, 1976). A short description of these systems is given showing their originality among software tools and their evolution; then two examples of expert system generators are presented to describe their specific features and their complementarity.

9.2.1.1 Main components. The first characteristic of expert systems is the separation of knowledge in an application domain (even limited) and the way this knowledge is used (the control structures).

Expert systems are composed of three main parts: the data corresponding to a special case (or fact base), the knowledge base and the control structures or the kernel of the expert system.

The fact base: The data contained in the fact base are the input data of the current problem plus all the deduced data added during the reasoning. These data may be symbolic data, numerical data or more complex structures like for instance a list consisting of an attribute, a value and a coefficient of plausibility.

Example:

> (colour blue 0.8)
> (perimeter 122 1)

The data in a fact base are usually organised in lists:

Example:

> ((colour blue 0.8)(perimeter 122 1)...)

In recent systems the fact base is organized in a more structured way, made up of objects and subobjects with various slots; their values are filled with the input data or added during the reasoning:

Example:

> Name = NGCxxx

ring = unknown
bar = present (0.4), absent (0.6)
shape = spiral (1)
ellipticity = 0.2 (1)

The knowledge base: The knowledge bases in early systems were large sets of production (or IF-THEN) rules. Most of the time the order of the rules in the base was important and could change the reasoning.

Knowledge bases in recent systems are structured (automatically or otherwise) into small bases to enhance the efficiency. Experts or knowledge engineers can express their knowledge through various knowledge representation schemes (frames, semantic nets, production rules, procedural knowledge, scripts, etc.). Each of these schemes may be pertinent or well adapted to express different kinds of knowledge.

For instance:

• Frames are well adapted to describe in a declarative mode the components of an object (physical or abstract).

• Semantic nets are well adapted to express the relations between objects.

• Production rules are used to express, in a dynamic mode, heuristic criteria.

For galaxy classification, as will be shown in the following sections, we need several knowledge representation schemes to describe the objects (for instance the classes), their relations (hierarchical) and heuristic criteria (for example to diagnose the presence of special features).

The control structures: The control structures which work with the knowledge contained in the knowledge base are the kernel of the expert system because they define the type of reasoning which is performed by the expert system.

Among the possible kinds of reasoning that existing systems perform, one can mention diagnosis, planning, simulation, classification, conception or interpretation.

For galaxy classification we need of course classification reasoning to handle knowledge of galaxy classification but as we also have to process images we need another kind of reasoning (planning plus execution control reasoning) to handle knowledge of the processing of images of galaxies.

9.2.1.2 Important features of expert systems. As has been shown in the previous section, expert systems enable symbolic data (as well as numerical data) to be handled and they perform logical reasoning. Another important advantage of expert systems is that they are good development tools for building complex systems. In fact they provide explanations, which is interesting both for a non-specialist to understand the logical reasoning but also for the expert developing a large knowledge base. As the knowledge is separated from the control structure (leading to a modular and explicit system) modifications in the knowledge base are immediate and easy to perform. Moreover, some extensions of the knowledge base can be added step by step, enabling the building of a huge system such as a real galaxy classification system, enhancing progressively its power and sophistication but keeping it coherent. Efficiency of the expert system is preserved by the automatic translation of the knowledge base into compact code using software engineering techniques (parsing, compilation, etc.).

9.2.2 CLASSIC

In this section we present the main characteristics of CLASSIC (Granger and de Mongareuil, 1987), an expert system generator designed for pattern recognition and diagnosis using classification reasoning. This system, developed by ILOG, was initially built at INRIA for galaxy classification; since then it has been used for several other application domains (Pouget et al., 1988; Thonnat and Gandelin, 1988). The main characteristics of CLASSIC are:

 • the organization of knowledge in description trees (called prototype trees) to explicitly describe the objects to be classified;

 • the use of deductive rules to represent dynamic or heuristic knowledge;

 • and the treatment of both imprecise and uncertain knowledge.

In addition, in CLASSIC there is:

 • a highly interactive development system including graphic stepping and debugging;

 • simple integration for existing programs written in C or Fortran;

 • and the structuring and automatic compilation of knowledge guaranteeing modularity and efficiency.

9.2.2.1 The Knowledge Base. The knowledge base is made up of three main components: the prototypes, the tree of prototypes and the rules.

The prototypes: In CLASSIC a portion of the knowledge is described by prototypes. The prototypes are typical descriptions capable of representing, for instance, physical objects, events or failures.

For example here is the prototype representation of a tiger (an example taken from the animal classification system of P.H. Winston, 1977):

Example of prototype:

        Prototype Tiger
        is-a: carnivore
        teeth: pointed
        pattern: striped
        colour: (yellow black)

This kind of concept was traditionally represented by production rules:

Same example using a production rule:

        If the animal is a carnivore
          and animal has stripes
          and animal has pointed teeth
        Then animal is a tiger

By describing the animal with a prototype, there is an explicit description of the concept. In addition it is declarative, avoiding the procedural and dynamic aspect of production rules. Moreover rules have the disadvantage that their fine granularity requires a single concept to be split up into several rules.

The prototype trees: In order to better organize the knowledge, prototypes can be linked to form a specialization tree going from the most general descriptions to the most precise. In our preceding example, the Tiger prototype belongs to the more general carnivore class. Such an organization fulfils two objectives:

• structuring of the knowledge: due to the inheritance of descriptions from fathers to sons, redundancy is eliminated and modularity is enhanced;

• control of reasoning: the hierarchical organization of the knowledge defines natural reasoning steps and permits efficient control.

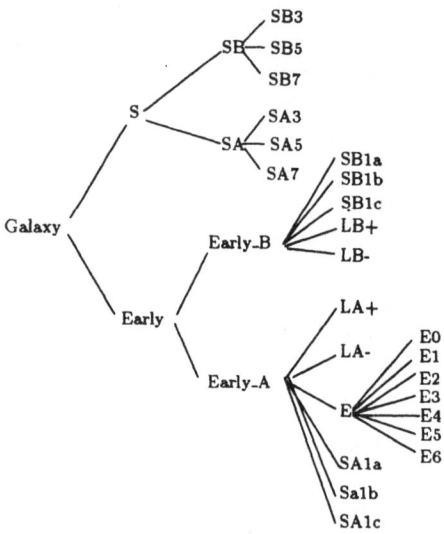

Fig. 9.1. A tree of prototypes.

The deductive rules: As well as the explicit representation of concepts, it is necessary to manipulate other kinds of knowledge by rules. In CLASSIC, rules are essential to allow description of the following knowledge:

• transformation of numerical information to symbolic information:

  if temperature > 38 then fever = high

  if length > width ×10 then shape = elongated

• synthetic knowledge:

  if shape of contour1 member of (average, spiral)

  and shape of contour2 = average

  then arms of galaxy = visible

CLASSIC uses these rules to add information when beginning with initial data.

9.2.2.2 The control structures.

Action of the engine: The inference engine in CLASSIC traverses the prototype tree searching for one or many solutions. At each node in the tree, CLASSIC uses the following basic cycle:

1. collect an initial set of information about the case to be handled;

2. use the production rules to infer as much knowledge as possible from this information;

3. choose the prototype most compatible with the available information;

4. attempt to confirm the choice of prototype with possible additional information;

5. continue toward the children of the prototype if this is confirmed, otherwise choose the most compatible prototype.

This process continues until one or all of the solutions are found.

Treatment of imprecision and uncertainty: CLASSIC is capable of representing and manipulating imprecise and uncertain knowledge using results from possibility theory and fuzzy set theory.

• The imprecise fact "fever is high" is characterized by a possibility distribution which corresponds to the characteristic function of a fuzzy set (for example on the interval in degrees Celsius, [38,39])

• The uncertain fact "John seems sick" is characterized by a confidence factor (called a possibility measure) and a doubt factor (which is the certainty of the negation).

Current expert system generators typically manipulate uncertain information by way of certainty factors; but they rarely handle imprecise information.

The use of possibility theory has permitted the defining of mechanisms to match rule premises with facts or to match prototypes with the object under consideration, neither of which is an "all or nothing" process.

Rules in CLASSIC can contain fuzzy predicates like $\sim=$, $\sim<$, *close to, far from* which permit the manipulation of imprecise numbers. For example, the pattern matching of the premise *"If temperature close to 39 degrees"* with the fact *"temperature = 38.8 degrees"* returns the possibility measure that 38.8 belongs to the fuzzy set defined below with value 0.5:

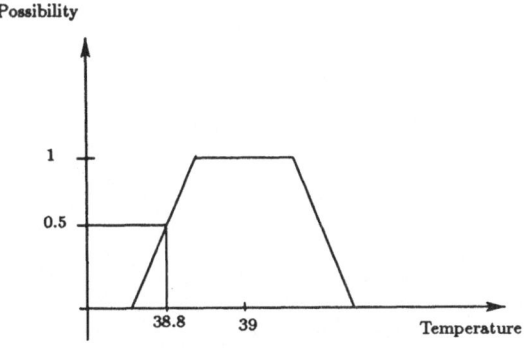

Fig. 9.2. The fuzzy predicate *close to.*

Control Slots: Control of the search for solutions in the tree is described by specialized slots of the prototype. These slots are used to describe the control of the process of reasoning. Thus, there is no parasitism of the application by introducing control information in the rules or mixing control rules and domain rules. The control slots specify:

• The number of solutions desired. For certain problems or parts of problems, one or all acceptable solutions may be required.

• Thresholds indicating the limits of acceptability of a solution.

• The importance of a prototype slot relative to others. This information from the knowledge of the expert, represents importance in the domain.

• The way to determine components of the current object.

• The order in which the slots of an object must be used and the initial data required once an object is selected.

• The rules to apply once a prototype has been accepted or rejected.

These control slots are automatically filled by CLASSIC with default values. They are constantly editable with the prototype editor.

9.2.3 OCAPI

OCAPI, which is currently being developed at INRIA for vision applications (Thonnat and Clément, 1988; 1989), is an expert systems generator specialized in the monitoring of libraries of procedures.

The main characteristics of OCAPI are:

• the structuring of knowledge into frames describing the actions performed by the library;

• the use of semantic nets to make explicit the decomposition of processing into several steps;

• the use of specialized production rules (choice between methods, evaluation of results or adjustment of parameters);

• logic of planning (generation of a list of actions to perform) plus control of execution (adjustment of the input parameters and evaluation of the results): see Fig. 9.3.

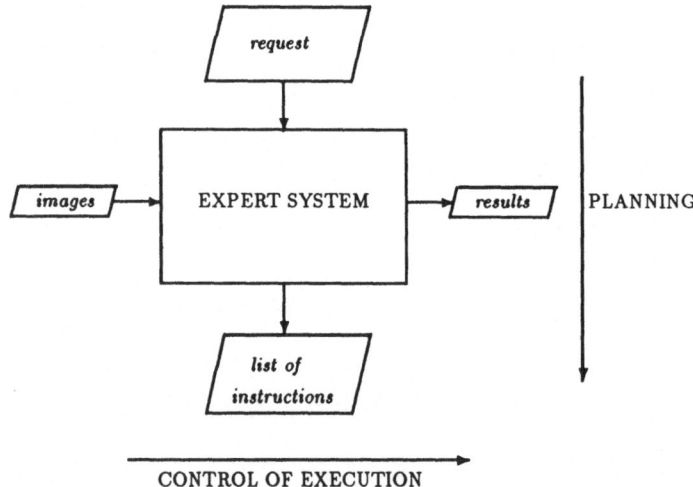

Fig. 9.3. The reasoning.

Some other features are:

• utilization in two modes: automatic or interactive;

• effective execution of binary code of the library;

• the possibility of monitoring procedures written in C , Fortran or Lisp.

9.2.3.1 The knowledge base. The knowledge base is made up of goals, operators and requests.

The goals: Each goal represents an image processing functionality; it is implemented as a frame. The main slots of the object are: the functionality, the input and output parameters (each one is a frame), and the knowledge to choose between methods and to test the results (implemented with rules: choice rules and evaluation rules).

Example of a goal:

Goal Filter1
    functionality:
                filtering
    input-parameters:
                { noisy-image: type image ... }
    output-parameters:
                { filtered-image: type image ... }
    choice-rules:
                { R1: if noise important
                    then use recursive-gaussian-filtering
                R15: if noisy-image binary
                    then use morphological-filtering
                R17: if slopes important
                    then use median-filtering }
    evaluation-rules:
                { R3: if filtered-image variance > request variance
                    then noise too-high }

The operators: Each operator knows how to reach a goal; it is implemented with a frame. The general slots are: the functionality, the input and output parameters, the characteristics which make explicit the effects of the operator (in terms of the parameters of the goal), and the adjustment which expresses the way to modify the result by changing the values of the input parameters (implemented with rules).

These operators are divided into two subclasses: the elementary or primitive operators and the complex ones.

Primitive operators: These correspond to a program of the library. They inherit the general slots of the operators but they have their particular slots such as syntax and programming language, to describe how to run the procedure.

Example of a primitive operator:

Operator med
    functionality:
                filtering
    characteristics:
                { median-filtering }
    input-parameters:
                { noisy-image: type image;
                  ws: type integer }
    output-parameters:
                { filtered-image: type image ... }
    adjustment-rules:
                { R8: if noise too high
                    then increase ws
                R11: if $\exists S, \forall object$ object size > S

then ws $< 2 \times S$ }
syntax: med noisy-image filtered-image -s ws

Example of a parameter described by its main slots:

In-parameter ws
        type: integer
        range: [1, 15]
        default: 3
        function: window-size

Complex operators: These correspond to a particular decomposition of a goal into steps. Each step is a special request of a subgoal. The decomposition is implemented with a semantic net where the nodes are requests and the links are *and* or *then* links.

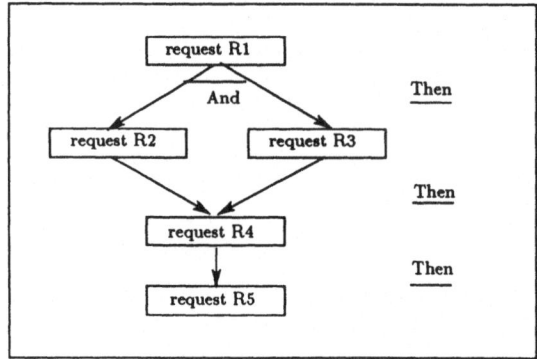

Fig. 9.4. A decomposition of a complex operator.

The requests: The knowledge found in the goals and the operators will be used to satisfy requests, coming from the user or the system. Requests are goals to be reached in a certain restricted context. They are implemented with frames. Their main slots are: the goal and the characteristics of the context, which are the restrictions on the input parameters and the specifications on the results (output parameters of the goal).

Example of a request:

Request R4
      median-filtering with *window-size* < 5

9.2.3.2 The control mechanisms. The role of the OCAPI system is to run programs to satisfy requests; i.e. to reach a goal within a certain context. The control mechanisms are built around two tasks: planning and control of execution, each control mechanism using for that the appropriate knowledge found in the knowledge base.

Planning: Given goals expressed with requests, the role of planning is to find the operators (the plan) to satisfy the requests.

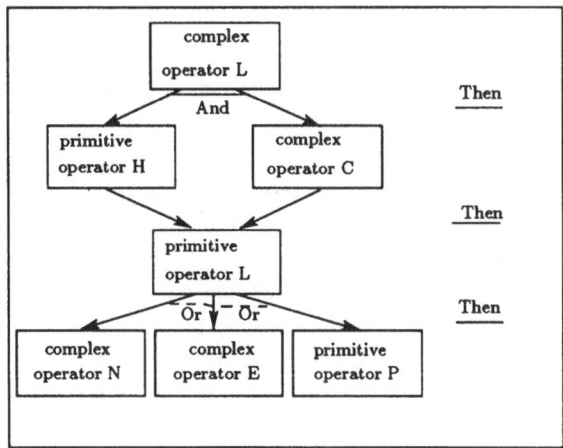

Fig. 9.5. A plan.

Planning proceeds in two steps: a simple selection phase and a more complex phase using inference.

The Control of Execution: *Control of execution* tries to execute the plan in such a way that the specifications on the results expressed in the request are satisfied. Following the choice of the next operator to execute and the initialization of each of its input parameters, control will proceed next to the effective execution of the operator, according to its type; *for a primitive operator:* the control mechanism calls the binary code of the library procedure; *for a complex operator:* the control mechanism decomposes it into less complex requests. Then, *the evaluation of the results* is achieved automatically according to the specifications of the request or by graphic presentation of the results to the user who evaluates them interactively. In case of unsatisfactory results, another execution will be performed after adjustment of the input parameters of the operator or following the choice of another operator.

9.3 The Knowledge to be Expressed

A really "intelligent" system in galaxy classification must handle several kinds of knowledge just as the astronomer does. Firstly, classification must take into account the nature of the context of the observations and of the formation of the images. Secondly, an automatic system has to process the data contained in the images using expertise in data reduction (preprocessing and object description). Then the complete description of the processed galaxy is matched with the classes using the knowledge of the description of the 3-dimensional models of galaxies.

For each of these kinds of knowledge (context, preprocessing, object description, class description) we detail the various elements of their contents, their use and where relevant how they can be implemented into knowledge representation schemes.

9.3.1 The Context

9.3.1.1 Contents. The elements of this knowledge are:

– the observed region (is it a region with special objects: cluster, resolved stars, and so on?);

– the observation instrument (is it a large field instrument: Schmidt camera or otherwise, what is the resolution, the quality of these instruments in general, of this particular instrument?);

– the physical device (what kind of device is it: a photographical plate, CCD, etc., if a CCD what are its specifications?);

– the wavelength (is it a special wavelength: H_α, V, IR, etc., or a set of them, how are they to be processed, to be interpreted, or to be mixed?);

– the exposure time (the value in minutes, the symbolic value: short, normal or long, or the knowledge to deduce the symbolic value from the numerical one);

– the observational conditions (seeing, sky background and absorption expressed with a scale of qualities: poor, average, good, exceptional);

– the digitizer (the type of digitizer: microdensitometer, CCD, Image Photon Counting System, etc.; the characteristics: noise, resolution);

– the calibration (presence of data, type of calibration: receiver, astronomical, etc.).

9.3.1.2 Usage. Essentially the context is used as a filter.

1. A filter to check if a value belongs to an interval or a set of predefined values.

Example:

Wavelength ∈ { B, V, R }

This feature allows one to make explicit the actual stage of development and skill of the knowledge base;

2. A filter to choose between several processings.

Example:

if calibration-data present
and data-type receiver-calibration
and receiver-type photographical plate
then execute calibration-chain photo

3. A filter to choose between several thresholds in an inference rule.

Example:

if exposure-time short
and ellipse-error > 0.4

then contour-shape spiral
if exposure-time normal
and ellipse-error > 0.2
then contour-shape spiral

In fact such knowledge (if it is supposed that this example is pertinent) has to be split into:
- context rules to define the threshold:

    if exposure-time short
    then err-ellipse-thr = 0.4
    and
    if exposure-time normal
    then err-ellipse-thr = 0.2

- description rules to use the threshold:

    if ellipse-error > err-ellipse-thr
    then contour-shape spiral

4. An evaluation of the quality of the input data and thus a way to measure the quality of further inferences.

Example:

    if observational-conditions ∈ { 2, 3 }
    and telescope = Mount Palomar Schmidt
    then validity-data correct

5. Finally the context data are useful for performing multisensor data analysis. This filter does not work on the values of the data but on the presence of certain types of information.

Example:

    if ∃ data-wavelength in B
    and ∃ data-wavelength in R
    then stars-age computable

9.3.2 The Image Processing

9.3.2.1 Contents. It is somewhat more arbitrary to separate from possible data processing approaches those which belong to preprocessing and those which belong to the object description. Moreover all this kind of knowledge has the same nature. According to a goal to be reached (description of images of galaxies), several steps have to be defined and the available and pertinent procedures have to be selected, implemented and controlled. We use for this part of the knowledge the OCAPI notions of planning and control of execution (see section 9.2.3) or more generally knowledge of automatic monitoring of procedures.

Descriptive and relational knowledge: The first type of knowledge to describe here is the decomposition of the image processing into the steps which are always required:

– sky background reduction (background mapping),
– noise reduction,

- galaxy detection,
- galaxy location (centre of the bulge if visible),
- galaxy isolation (detectable limits of the galaxy),
- star removal (and the removal of all objects except the galaxy),
- galaxy description,
- galaxy classification.

These steps represent goals to reach; so one operator, *galaxy-processing*, could be implemented using the OCAPI terminology like this:

Example:

Operator Gal-proc1
    functionality:
                     galaxy-processing
    characteristics:
                     { unique-input-image }
    input-parameters:
                     { input-image: type image;
                       context: type frame }
    output-parameters:
                     { class: type symbolic }
    decomposition:{
                           sky-background-reduction
           then
                           noise-reduction
           then
                           galaxy-detection
           then
                           galaxy-location
           then
                           galaxy-isolation
           then
                           star-removal
           then
                           galaxy-description
           then
                           galaxy-classification
           }

Descriptive and heuristic knowledge: In fact, for the same functionality (or goal), i.e. galaxy classification, several operators can be implemented according to the nature of the observed images (one, two or several images) and of the calibration data.

Among the steps of the decomposition of the operator some can be quite complex; for instance star removal can be performed using various methods: by the shape (correlation with the point spread function), by the intensity (detection of the bright objects), by the colour (if the wavelength is different from those of the HII regions), or by the position (if some stars are identified in the field).

Such knowledge can be implemented using production rules; these rules are the choice rules attached to the goal *star-removal*:

Example:

Goal Star-removal
    functionality:
                     star-removal
    input-parameters:
                     { spoiled-image: type image ... }

output-parameters:
{ cleaned-image: type image ... }

choice-rules:
{ R1: if ∃ identified objects
 then use remove-by-position-and-shape
 R2: if wavelength $= H_\alpha$
 then use remove-by-intensity-and-shape
 Rn: if ....
 then use ... }

For each of these steps we need to have a good description of all the available operators and of the way to choose between them; since some of them can be subdivided into distinct parts we can still have complex operators for different levels of abstraction.

For instance an operator which performs galaxy description can be implemented as a complex operator with a decomposition into a first step of photometric description and a second step of morphological description.

To automate image processing we also need to introduce knowledge of the evaluation of results for various levels of requests to subgoals; this can be done using heuristic criteria implemented with rules attached to the goals. For instance if we suppose that the goal *galaxy-detection* provides as output a segmented image and a measure of the size of the detected object, a heuristic rule can evaluate the quality of the detection (which depends on thresholds difficult to initialize) by testing the size of the detected object.

Example of goal with evaluation rule:

Goal Galaxydetection
 functionality:
  galaxy-detection
 input-parameters:
  { complex-image: type image ... }
 output-parameters:
  { detected-object: type frame }
 evaluation-rules:
  { R3: if object size ≤ star size
   then detected-object ambiguous }

Once the request has failed we need to use some knowledge to readapt the processing according to this evaluation. This can be done using heuristic criteria attached to the operators available to perform the desired goal. For instance if the galaxy detection is performed by the operator Galdet1, adjustment rules can express how to modify the input parameter in such a situation:

Example of an operator with adjustment rule:

Operator Galdet1
 functionality:
  galaxy-detection
 input-parameters:
  { complex-image: type image ... }
 output-parameters:
  { detected-object: type frame }
 adjustment-rules:
  { R1: if detected-object ambiguous
   then less strict thresholding }
 decomposition:
  { ....

then
                                    thresholding
then
           .... }

In one of the operators performing the thresholding we know through other adjustment rules how a less strict thresholding can be obtained:

Example of an operator with adjustment rule:

    Operator Muls
        functionality:
                thresholding
        input-parameters:
                { greylevel-image: type image
                thr: type float ... }
        output-parameters:
                { binary-image: type image }
        adjustment-rules:
                { R19: if less strict thresholding
                    then increase thr }

9.3.2.2 Steps. Now we detail the various steps.

Sky background reduction: The determination of the primitives for the morphological analysis is very sensitive to the sky background reduction. Many techniques have been proposed to remove the background.

One of the first such techniques, numerical mapping (Jones et al., 1967), is based on an iterative adjustment of the data with a second degree homogeneous polynomial. At each step, we reject the pixels for which the deviation is greater than 3 times the standard deviation. After some steps (3 to 4) we obtain a clean map.

Some improvements were achieved using a division of the image into blocks. In each block a first background value is determined using the intensity histogram (modal, median or other estimator). This leads to fast convergence. The spline functions (B-splines, pseudo or true splines) are often used to compute the map, after the determination of the background in each block (Bijaoui, 1980; Kurtz et al., 1985).

In the mapping, the galaxy region is removed, in order to suppress the influence of the galaxy on the background mapping.

Other methods are also used, based on star removal and cicatrisation (Thonnat and Llebaria, 1983). Some systematic errors could result from bad determination of the background in the wings of the objects.

Noise reduction: Morphological analysis and parameter extraction depend on the noise reduction carried out on the image. Generally, classical smoothing with a gaussian profile allows the best compromise to be obtained. The gaussian profile is chosen as being similar to the stellar one. This smoothing conserves flux but modifies some structural parameters such as central elongation and gradients. The median filter, on the other hand, does not conserve flux, but preserves the mean slopes. This filter may suppress important structural features, like hot spots or the galaxy nucleus. Other enhancements, such as non linear deconvolution (Titterington, 1985) could be used, in order to improve the detection of arms, bars, rings or other kinds of pattern.

Galaxy detection: Bright galaxies are known for many years and therefore it is easy to locate their positions using a computerised catalogue. There may be some problems to locate them automatically on a large digital image. Taking into account the background mapping and the classical detection technique we often detect them as a set of objects, and not as a whole.

For fainter galaxies, the identification strategy is very different (MacGillivray and Stobie, 1984; Kurtz et al., 1985; Slezak et al., 1988). After classical pre-processing, a detection threshold is determined. The objects are extracted using image segmentation, allowing all connected fields to be built up. Some primitives are computed at this stage: area, integrated flux, barycentral position, various moments or extrema.

Separation of galaxies from stars is performed by Bayesian classifiers. The area, the integrated magnitude, and a concentration parameter derived from the weighted moments are the variables used generally in the analysis.

It is evident that detection based only on appearance eliminates compact objects. Only a careful photometric analysis, spectrographic or radio observations can lead to the detection of galaxies with a quasi-stellar shape.

Galaxy location: Image segmentation (Rosenfeld, 1969) leads to the approximate location of the galaxy. When the galaxy is very large, the centre obtained by this technique (weighted or otherwise) may be far from the real centre.

Determination of the intensity maximum gives another position which can be better than the barycentre, but can also be very far than the real centre (due to a superimposed object, a hot spot, etc.)

The centre can be determined using a set of rules, taking into account distance to an isophote, or the relative intensities of the internal maxima.

Multi-threshold image segmentation can give also some information on the structure of the galaxy, giving an explanation for the divergence between barycentre and maximum multiple object, fuzzy centre, etc.

Limits of the galaxy: Morphological analysis depends strongly on the limits given to the galaxy.

The threshold at a given level never gives good limits: if the threshold corresponds to a magnitude of $23"^2$, it is too high and the photometric parameters are underestimated; for a magnitude of $25"^2$, the noise is too high to obtain a good contour, and the morphological parameters are badly estimated.

A good method is derived from the radial profile (Le Fèvre et al., 1986). Accuracy increases with the distance, and we can determine the limit at a high magnitude, such as $26.5"^2$.

In many cases, particularly in clusters of galaxies, two galaxies can be so close that one galaxy contaminates study of the other. Very few investigations have been published on the most appropriate way to process the corresponding images. It is evident that we cannot remove a galaxy without good knowledge of the profile.

Star removal: After detection and determination of the limits of the galaxy, it is necessary to remove stars and possible superimposed objects.

Identification: It is not evident how to identify which objects belong to the galaxy, and which can be considered as field stars. A stellar-like object can be a faint globular cluster, an OB association or a superassociation, a HII region, a hot spot, a supernova, etc. In each of these cases, the presence of these objects can yield important information as regards galaxy classification and so we need to try to eliminate only the features external to the galaxy, essentially galactic stars.

Position of the objects is the first criterion. A bright object far from the galaxy centre is generally a superimposed star. Photometric analysis may roughly separate normal galactic stars from some features belonging to the galaxy studied (blue associations, globular clusters, etc.). Spectrograph observation, giving the radial velocity, is the perfect tool to indicate which objects belong to the galaxy or do not. Some observations with narrow-band filters, or scanning Perot-Fabry interferometers, may also yield this information.

Study of light variabilities, polarisation, or some radio spectroscopy or radio mapping allow us sometimes to obtain a good criterion.

Number of objects to remove: The number of objects depends on many parameters: galaxy size, limiting magnitude, celestial position, etc. The position, with regard to galactic plane and galactic longitude, is an essential parameter. If the galaxy is high on the galactic plane absorption and stellar density are small; on the other hand, it is very difficult to detect and to identify a galaxy in the galactic plane, especially in a direction close to the Galaxy centre.

The number of objects to remove is typically between ten and thirty.

Reduction technique: Various techniques have been proposed to remove point-like or extended objects.

For stellar shape objects, the theoretically best technique is based on a local adjustment of a stellar profile. Then the resulting profile is subtracted. The residues are around 3 magnitudes higher than the surface magnitude before subtraction (Debray, 1988).

Another classical method uses cicatrisation (Thonnat, 1982). The object to be removed is circled, and the internal field is filled step by step, using linear interpolation with the values of the previous contour. Some small defects are observed, especially a smoother internal texture. This tool can be applied to point-like or extended objects.

Faint stars, or some artifacts, can be removed using a classical median filter (Thonnat and Llebaria, 1981). This operation does not conserve flux, but the morphology is not really modified. Another advantage of this filter is to modify the values only on a small scale.

In data analysis in the future, it may be necessary to examine the influence of the removed regions on the computed parameters.

9.3.3 Galaxy Description

9.3.3.1 Galaxies as physical objects. The identification of the real nature of the galaxies (Hubble, 1936) begins in the 1920s with the first spectrographic observations.

Generally a galaxy is not an isolated object in the Universe; it is a concentration of matter included in a group or a cluster. It is clear that the clusters themselves are correlated, and there exist superclusters and other huge structures (de Vaucouleurs, 1970; Peebles, 1980; Oort, 1984). But why do so many galactic patterns exist?

The first fundamental parameter is the total mass. Classical elliptical galaxies are more massive than spiral ones (Binggeli et al., 1988). Irregular galaxies are generally less massive objects, as are dwarf ellipticals. The mass is not the only parameter which determines morphology. A galaxy is a mixture of stars, gas, dust and probably non baryonic dark matter (Faber, 1987). Morphology is strongly correlated with the distribution of matter.

A normal or a dwarf elliptical galaxy has only a small percentage of matter in the form of gas or dust. Almost all the luminosity is due to old stars. On the other hand, spiral galaxies have a large fraction of matter in the form of gas and dust. There is evidence in this case for newly formed stars. Irregular galaxies also show recent star formation.

The following scheme tries to explain the structure of galaxies (Freeman, 1975; Silk, 1986). A large mass of matter collapses (gravothermal instability) and in the core a first population of stars is formed, embedded in a large halo of gas. The stars evolve, the biggest rapidly, the smallest very slowly. There is an equilibrium between the gas ejected from the stars and the formation of the new ones. The galaxy environment plays a very important part in the gas content, and the evolution of the mass (Silk, 1986). Many phenomena can be identified: gas stripping by a close encounter, galaxy merging, tidal interactions, cannibalism of small galaxies by giant ones, etc. If the gas rate is too small, no star formation takes place, the population becomes aged, is mixed and we have an elliptical object; if the gas rate is high stellar formation is possible. Generally stars are formed by the interaction of a density wave with the gas. This perturbation has a spiral shape (Lin et al., 1969). Other perturbations are possible leading to irregularly distributed, newly formed stars.

Studies of morphological types in clusters of galaxies have proved the importance of environment (Dressler, 1980): elliptical galaxies are in dense regions while spiral galaxies are in sparse ones. Almost all curious morphological objects have been explained by encounters between two or more objects.

Another phenomenon plays an important role: nucleus activity (Sersic, 1982). Classification has been carried out only on its appearance. Photometric and spectroscopic observations have shown the real importance of the nucleus even for normal galaxies. Its properties are not completely understood in the classical scheme. The most active nuclei are always associated with giant elliptical galaxies (when the galaxy can be seen).

The light distribution results from the stellar dynamic (Freeman, 1975). N-body computations have been used to try to explain different distributions (Martinet, 1985). For elliptical galaxies, a $r^{\frac{1}{4}}$ law (de Vaucouleurs, 1948) is observed. It is interpreted as a truncated isothermal distribution. S0 galaxies show a mean exponential decreasing law (Johnson, 1961). For many galaxies, we can have a superposition of two laws, corresponding to two different populations (disk and bulge; Simien and de Vaucouleurs, 1986).

Morphological classification must take into account all this knowledge. The choice of primitives is guided by the interpretation: if the object is smooth, it is probably elliptical; if it is granular, this is due to stellar formation and the galaxy can be spiral or irregular. If the profile decreases rapidly, it is a spiral; if it follows the $r^{\frac{1}{4}}$ law it is elliptical, etc.

9.3.3.2 Photometric measures.

Profiles: The radial profile is the main photometric characteristic. It is easy to estimate a profile, but its quality depends (i) on the manner in which the stars have been removed, and (ii) on the way the discrete geometry of the digital image is taken into account. Spline interpolation may reduce the sampling effects. Fuzzy integration (Bijaoui, 1988) permits also realistic profiles to be computed, without image extension.

The profile can be computed in elliptical rings, in order to have a quite uniform value in each ring. Axis ratio and orientation are computed on a given typical isophote.

The profile projected on the major and the minor axes can also be estimated easily (Watanabe et al., 1982).

Parameters derived from profiles: The radius r_l from a given magnitude m_l is easy to obtain, when the field is not too crowded.

Together with the determination of the radius, we compute the integrated magnitude m_{il} for a given threshold. Surface brightness m_{sl} is obtained by:

$$m_{sl} = m_{il} + 5log_{10}(r_l) + 2.5log\pi$$

Surface brightness is an important parameter in galaxy classification.

Total magnitude m_t is estimated by the study of the profile convergence.

It is interesting to obtain the magnitudes at given radius m_r in order to compute the colour indices.

The profile is also a characteristic of morphological type. For elliptical galaxies, the profile follows de Vaucouleurs' law:

$$m = ar^{\frac{1}{4}} + b.$$

For S0 galaxies, the mean profile follows an exponential law:

$$m = ar + b.$$

The parameters a and b are related to the equivalent radius r_e and the equivalent magnitude m_e. They can be computed with the least mean square estimator.

Other laws are often used like Hubble's or King's profiles (Holmberg, 1975; Kormendy, 1980).

The central part of the profile is difficult to correctly estimate. Deviation from de Vaucouleurs' law may be interpreted as the existence of a massive central attractor (Kormendy, 1980).

Bulge/disk ratio: For spiral and lenticular galaxies, a classical approach assumes superposition of two populations. The older one, the bulge population, follows de Vaucouleurs' law, while the younger one corresponds to an exponential decreasing law.

The profile can be decomposed into two parts (Simien and de Vaucouleurs, 1986). The ratio between the two populations gives another classification criterion.

Generally the bulge is nearly round, while the disk is not seen face-on. The ratio between the populations depends on the angular position. Two profiles are computed in separate regions, taking into account the axis of the disk population.

Photometric indices: Galaxy evolution can quite perfectly explain the value of the colour indices (Rocca-Volmerange, 1985). Roughly, elliptical galaxies are red because the old red stars give the main part of the flux, while the newly formed bright stars give rise to a blue index for spiral galaxies.

It is important to estimate the index in the same area, and not for the same limiting magnitude.

Flux/area measures: We have seen that we derive a surface magnitude from the radial profile. We can obtain a similar parameter from the integration on the field limited to a given isophote. The difference is small for regular elliptical galaxies, while the result can be different from spiral or irregular ones.

9.3.3.3 Morphological parameters. Different morphological features can be extracted from the images.

First of all, some global parameters have to be computed to describe the morphology of the galaxy (orientation of the main axis, elongation, surface areas, perimeter, first moments; Jarvis and Tyson, 1981; Grosbøl, 1987; Di Gesù and Maccarone, 1986; Accomazzi et al., 1989).

This description can be completed by the detection of other special features (such as texture, presence of bar, arms, etc.).

Texture: Description of the texture is very important for a good classification. It can be a simple coarse distinction between a highly smooth texture or otherwise (computed using variance only) but even for a simple measure we need to take care of the quality of the noise reduction. A finer texture description would be useful (but it has not yet been carried out) to describe the nature of inhomogeneities in the disk; this texture is of course related to morphological type but also to resolution.

Bar: Several approaches can be used to detect the bar which is a bright elongated feature across the bulge (Thonnat, 1985). For some of these approaches, we need to measure elongation for several isophotes representing the various regions in the galaxy. These elongations have to be compared with the global elongation of the galaxy. Another approach for detecting the presence of the bar is to measure the orientations of several isophotes and to compare them with the global orientation of the galaxy.

Arms: Generally, arms have very complex shapes, so the isophotes describing the various regions in the galaxy have shapes which are far from elliptical. A measure for each isophote of the distance from the best ellipse, together with compactness, leads to an evaluation of the presence of a spiral shape (Thonnat, 1985). But much work still remains to be done in order to describe the structure of the arms. For this, ridge detectors (Fairfield, 1989) adapted to galaxy images are currently being studied.

Ring: For bright galaxies the description of morphology requires the detection of possible internal or external rings. These features are not easy to isolate; they can be circular, elliptical as a function of orientation, or with an "8" shape; they can be perturbed by proximity of the arms, or even disrupted. At the present time, they are essentially visually detected and described (Buta, 1988). For automated detection, peak detector algorithms have to be implemented.

Symmetries: Another important characteristic, easy to measure, is the relative symmetry of the shape or of some subparts (main axis profile asymmetries). This is a good way to detect irregular galaxies (magellanic galaxies).

9.3.4 The Classification System

The goal of this section is not to describe various galaxy classification systems, but to discuss the knowledge of the classification system that has to be expressed in the knowledge base.

9.3.4.1 Descriptive knowledge. The first kind of knowledge to be expressed is the description of the various classes: what is a spiral galaxy, what are the various subparts and how are they to be characterised? This knowledge is declarative, and can be represented using structured objects (frames).

Example:

CLASS: spiral
BULGE: { SHAPE: spherical
          TEXTURE: homogeneous
          PHOTOMETRY: $r^{1/4}$
          LOCALISATION: central }
DISK: { SHAPE: flat
        TEXTURE: non-homogeneous
        PHOTOMETRY: linear
        LOCALISATION: .... }

.....

9.3.4.2 Relational knowledge. Of course this description can be complete and precise to a greater or to a lesser extent; thus it is useful to organise these frames showing the hierarchical links existing between the various classes. Schematically the class Galaxy can be subdivided into three subclasses: elliptical, lenticular and spiral; each of these classes can be subdivided into different subclasses, and so on. This knowledge can be easily represented using semantic nets with the classes as nodes and the hierarchical dependencies as links (see Fig. 9.6).

9.3.4.3 Deductive knowledge. This descriptive knowledge is not sufficient to be able to actually recognize the object detected in the image (or images) as a galaxy belonging to one of these classes. Another kind of more operational knowledge is needed to perform the connection between the numerical or symbolic parameters describing the object of interest and the features used in the description of the

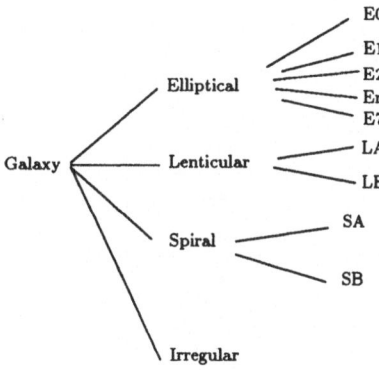

Fig. 9.6. A semantic net describing the various classes.

classes. This can be done with production rules expressing heuristic criteria such as:

if ellipse-error > 0.5
then contour-shape = distorted

if variance > 6
and contour-shape distorted
then texture inhomogeneous

9.4 Feasibility of Such Systems

9.4.1 Introduction

In this section we describe SYGAL, an attempt at building such a system. This work is described in Thonnat (1985; 1989). Figure 9.7 shows an overall summary of the system.

The system was developed for the automatic classification of images of bright galaxies taken with the Schmidt camera at CERGA (Centre d'Etudes et de Recherches Géodynamiques et Astronomiques, Grasse, France). The sampling rate of the images was (20 × 20) microns, and the image size was (512 × 512) pixels.

Among the properties of such images, we must note that these images have a very low signal-to-noise ratio, which becomes critical for faint objects. There is no clear limit discriminating the galaxy from the background. Other objects may be close, or even partially overlapping the galaxy.

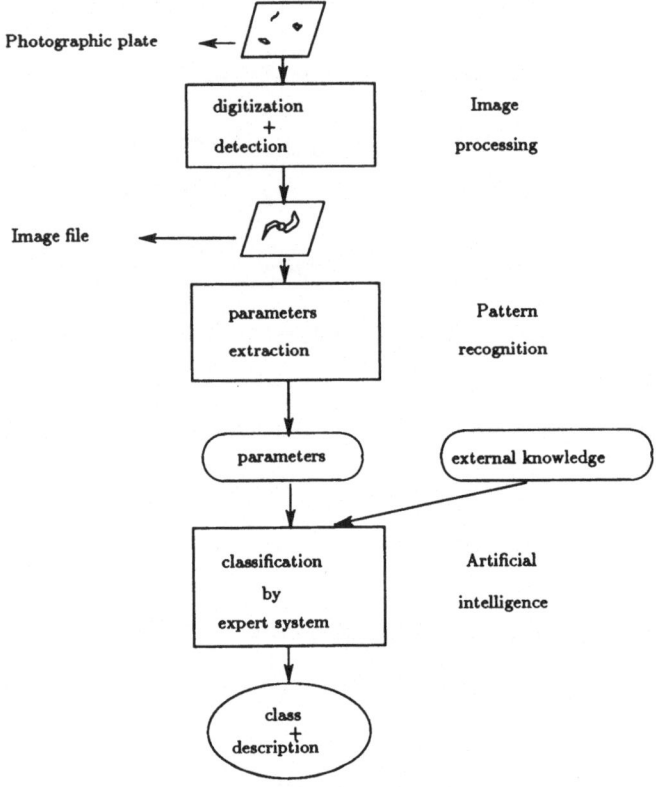

Fig. 9.7. Control structure of SYGAL.

9.4.2 Preprocessing

The preprocessing phase was a fully automatic phase which was performed purely algorithmically, i.e. in a deterministic manner and not using an expert system supervisor. This phase performed the following functions:

- location of the exact position of the galaxy;
- determination of the limits of the galaxy;
- building, and then suppressing, a luminosity relief map of the background;
- removal of objects in the neighbourhood;
- elimination of noise.

9.4.3 Parameter Extraction

In this section we present the various parameters extracted from galaxies to describe their shape.

9.4.3.1 Major axis. The first parameter computed is the orientation of the major axis of the galaxy. The value of its direction is given by the angle θ which minimizes the moment of inertia J:

$$J = \sum_{x,y} d(x,y)(y \cos \theta - x \sin \theta)^2.$$

Let θ_1 and θ_2 be the two solutions of

$$\tan(2\theta) = 2 \frac{\sum_x \sum_y d(x,y)xy}{\sum_x \sum_y d(x,y)(x^2 - y^2)}$$

with $J(\theta_1) < J(\theta_2)$.

- θ_1 is the orientation of the major axis;
- θ_2 is the orientation of the minor axis (orthogonal to the major axis).

A set of curves representing the intensity of the galaxy from the centre in different directions is built. These directions are the two principal half-axes and the two minor half-axes.

9.4.3.2 Ellipticity. The shape of the galaxy is very dependent on the viewpoint with which the object is observed. Therefore, an important measure is the apparent ellipticity of the galaxy: $e = 1 - \frac{b}{a}$ with a and b respectively the width of the major axis and the width of the orthogonal axis.

This value is provided by the previous curves indicating the width of the galaxy in the directions of the major and minor axes. As these curves are noisy and have a smooth slope due to diffusion in the emulsion, we use an integrated width value. Let $F(x)$ be the average distribution of the intensity along the major axis, and $f(x)$ the average distribution of the intensity along the minor axis; let w_1 and w_2 be respectively the width of the galaxy in the two directions.

The widths are deduced from:

$$\int_0^{w_1} F(x)dx = \frac{3}{4} \int_0^\infty F(x)dx$$

and

$$\int_0^{w_2} f(x)dx = \frac{3}{4} \int_0^\infty f(x)dx.$$

Finally the apparent ellipticity is given by: $e = 1 - \frac{w_2}{w_1}$.

9.4.3.3 Size. In order to estimate the reliability of the measured parameters we compute the size of the galaxy S with the hypothesis of an elliptical shape: $S = \pi w_1 w_2$.

9.4.3.4 Average profile. Using the measure of apparent ellipticity e we build the curve of the average distribution of intensity from the centre $g(x)$.

The curve $g(x)$ is the radial distribution of the intensity in the new referential direction $(X = x, /Y = \frac{w_1}{w_2}y)$ with X the coordinate on the major axis and Y the coordinate on the minor axis.

9.4.3.5 Projected profile. Theoretical studies have shown that the radial variation of luminosity from the centre of galaxies is the sum of two functions. The first function is the spheroidal component characterizing elliptical galaxies:

- the intensity I is given in terms of the radius r by $I_1(r) = a_1 e^{-b_1 r^{\frac{1}{4}}}$
- the density $d = \log(\frac{I}{I_0})$ is given in terms of the radius by $d_1(r) = \alpha_1 - \beta_1 r^{\frac{1}{4}}$.

The second function is the flat component characterizing the disk of lenticular and spiral galaxies:

- the intensity I is given in terms of the radius r by $I_2(r) = a_2 e^{-b_2 r}$.
- the density $d = \log(\frac{I}{I_0})$ is given in terms of the radius by $d_2(r) = \alpha_2 - \beta_2 r$.

In order to avoid the influence of the viewpoint with which the galaxy is observed we build a new curve: the projected profile d_p. This curve is obtained by orthogonal projection of the densities d of the galaxy along the major axis.

$$d_p(u) = \sum_{x,y} d(x,y)$$

with $u = x \cos\theta + y \sin\theta$, θ the direction of the major axis, x and y the coordinates of the pixels belonging to the galaxy. From this curve we extract two parameters:

- the parameter *profile* which is the ratio of the mean square errors made by approximating the projected profile respectively with d_1 and d_2 for r greater than the radius of the bulge. The estimation of the radius of the bulge is obtained from the average profile previously constructed:

$$profile = \frac{\sum_{r=r_{bulge}}^{r_{limit}} (d_p(r) - d_2(r))^2}{\sum_{r=r_{bulge}}^{r_{limit}} (d_p(r) - d_1(r))^2}$$

- the parameter *linear_err* measuring directly the mean square error made by approximating the complete projected profile for $r = 0$ to $r = r_{max}$ with a linear function:

$$linear\_err = \sum_{r=0}^{r_{limit}} (d_p(r) - d_1(r))^2.$$

9.4.3.6 Contours.

Contour building: In order to describe variation of structure in different regions of the galaxy, we need to extract several isophotes. The isophotes must be completely representative of each region in the galaxy and they must be built automatically. First we compute five thresholds from the distribution of the density along the major axis $F(x)$, then these thresholds are used to obtain five binary images on which we apply an edge detector algorithm.

The five thresholds $t_i, i \in [1, ..., 5]$ are given by:

$$\int_0^{t_i} F(x)dx = \frac{1}{n_i} \int_0^{\infty} F(x)dx$$

with n_i respectively equal to 0.20, 0.50, 0.75, 0.85 and 0.90.

For each thresholded image, we use a Sobel edge detector, then we examine the maximal chain resulting from the contour chaining algorithm of the INRIM-AGE library (Cipiere, 1984).

Parameters extracted from these contours: From each of these contours we extract five parameters characterizing its shape: ellipticity, angle of the major axis, relative position of the centre, compactness and distance to the closest ellipse.

Angle of the major axis: in the same way as for the image, we compute the angle θ_i which minimizes the inertial moment J_i and the orthogonal angle ψ_i.

Ellipticity: for each contour we compute the ellipticity (the description of this has already been given above).

Relative position of the centre: for each contour we compute the Euclidean distance (*centre_err*) between the centre of the bulge and the centre of gravity of the contour g_i.

Compactness: the compactness C measures the roundness of a shape and is minimal for a circle ($c \leq 4\pi$): $c = \frac{\text{perimeter}^2}{\text{area}}$. In fact we take into account the value of the estimated ellipticity $e = 1 - \frac{b}{a}$ and we measure the quantity: $C = \frac{b}{a}\frac{c}{e\pi}$.

Distance of the closest ellipse: the final parameter, *ellipse_err*, is the distance between each contour and the ellipse which has its major axis in the θ_i direction, ellipticity e and is centred at the centre of gravity g_i.

Normalization is performed as follows: let S_E be the area of the ellipse, S_C the area of the contour and S_{dif} the sum of the areas between the two closed curves; then:

$$ellipse\_err = \frac{S_{dif}}{S_E + S_C}.$$

S_{dif} is approximated by dividing the area into small triangles, the vertices of which are respectively, the points of the contour C_i, the projections of these points on the ellipse p_i and the intersections of the contour and the ellipse P_j. Let x_{p_i} and y_{p_i} be the coordinates of the projections p_i; and let x_i and y_i be the coordinates of the points C_i belonging to the contour; we have:

$$x_{p_i} = a\cos(\arctan(\frac{ay_i}{bx_i})) \quad \text{and} \quad y_{p_i} = b\sin(\arctan(\frac{ay_i}{bx_i})).$$

9.4.4 Example of Measured Parameters

We display the parameters extracted from NGC4569:

orientation:	79.33
ellipticity:	0.50
linear_err:	0.039
profile:	2.77
area:	38013.3

	contours	centre_err	ellipse_err	compactness	angle	ellipticity
c1:	1	0.06	1.6	−6.3	0.09
c2:	2	0.21	5.4	70.8	0.36
c3:	11	0.16	7.1	83.6	0.58
c4:	9	0.21	11.8	76.8	0.47
c5:	7	0.14	8.6	76.1	0.50

9.4.5 Automatic Classification

This phase was performed using an expert system.

9.4.5.1 The engine. The engine used for the expert system was the first version of CLASSIC (see section 9.2.2) which was built for this purpose. This version did not allow the call of external procedures (written in C or Fortran) so it was not possible to measure new parameters in the image during the reasoning. Some other interesting features such as a large number of weights (to balance the rules, the descriptors in the prototypes, and so on) and the multi-window interface were not yet provided in this version of CLASSIC.

9.4.5.2 The knowledge base.
   The symbolic parameters: The description of a class is made by the astronomer in terms of symbolic parameters which represent the different structural patterns of the galaxies. We have chosen to introduce such parameters in the knowledge base in addition to the quantified parameters previously described.
   The symbolic parameter H: this symbolic parameter corresponds to the hierarchical label coding; it takes the values of the possible classes:

$$H : E, L, S, E0, E1, \ldots, E6, LA, LB, LA-, LB-, LA+, LB+, SA, SB, SA0,$$

$$\ldots SA7, SB0, \ldots SB7.$$

As irregular galaxies are not yet classified the associated values are not available.
   The symbolic parameter T: this parameter corresponds to the continuous label coding; though it takes numerical values, it is not a measured parameter but rather each value represents a label:

$$T : -5, -3, -2, -1, 0, 1, \dots, 7.$$

Values greater than 7 are not considered since they correspond to irregular galaxies.

The symbolic parameter bar: this parameter specifies knowledge about the possible presence of a transversal bar; so, the symbolic values are:

$$bar : present, absent, unknown.$$

The symbolic parameter shape: this parameter qualifies the general structure of the galaxy; it may take the values:

$$shape : elliptical, average, spiral.$$

The symbolic parameter isophotes: this parameter points out the degree of perturbation of the contours; the possible values are:

$$isophotes : smooth, normal, distorted.$$

The symbolic parameter arms: this parameter characterizes the arms of the galaxy; the values are:

$$arms : absent, incipient, evident, branched.$$

The symbolic parameter bulge: this parameter is referred to if the central bulge can be detected on the projected profile; so, the values are:

$$bulge : visible, invisible.$$

The symbolic parameter flatness: this parameter characterizes the apparent ellipticity of the galaxy.

$$flatness : null, negligible, very\_faint, faint, light, average.$$

The symbolic parameter centring: the parameter centring measures the difference between the centres of the galaxy and the centre of the contours.

$$centring : good, average, mediocre, indifferent.$$

The symbolic parameter profile_concavity: this parameter indicates if the concavity of the curve of the projected profile is large or not.

$$profile\_concavity : great, average, null.$$

The symbolic parameter validity: this important parameter specifies the quality of the observation and thus the quality of the measured parameters; it is a function of the size of the image of the galaxy. Values are:

$$validity : good, bad.$$

We have seen in the previous section that for each processed galaxy five contours are built; these contours are described with both symbolic and quantified parameters. The quantified parameters correspond to the parameters which have been extracted from each contour: *centre_err, ellipse_err, compactness, angle, ellipticity.* Three symbolic parameters describe also the contours in the different regions of the galaxy.

The symbolic parameter contour c_i: shape: this parameter characterises the shape of each contour, in the same way as the symbolic parameter *shape*; it takes the same values as the global parameter:

$$\text{contour } c_i : \text{shape} : \textit{elliptical, average, spiral.}$$

The symbolic parameter contour c_i: isophotes: this parameter indicates the degree of perturbation of each contour; like the global parameter associated with the entire galaxy it can take the values:

$$\text{contour } c_i : \textit{isophotes} : \textit{smooth, normal, distorted.}$$

The symbolic parameter contour c_i: flatness: this parameter indicates if the ellipticity of each contour is not too large:

$$\text{contour } c_i : \text{flatness} : \textit{valid.}$$

The prototypes: There are 37 prototypes used to define the various classes; they are organized in a hierarchy which strictly reflects the hierarchy of the classes.

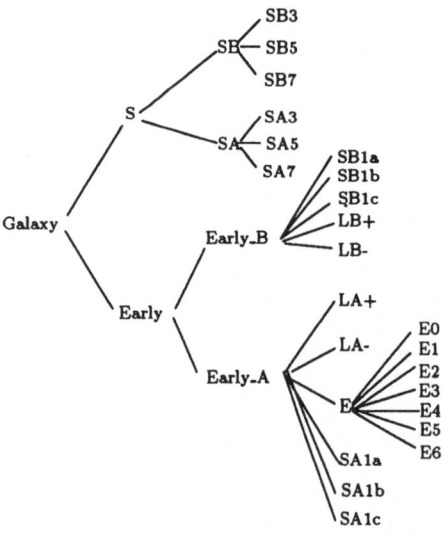

Fig. 9.8. The tree of the prototypes.

The descriptors of a prototype characterize the objects belonging to the corresponding class. Symbolic, numeric and complex structures are represented by different data structures. Some descriptors may have complex structures as shown in the following example:

Class Contour
compactness: [0, 10000];
ellipticity: [−0.9, 1];
angle: [−90, 90];
ellipse_err: [0, 10000];
centre_err: [0, 10000];
shape: elliptical, average, spiral;
isophotes: unknown;
flatness: unknown.

We show here some examples of prototypes describing the classes Galaxy, SB7 and E0.

Class Galaxy
T: [−5, 10];
profile: [0,10000];
ellipticity: [−0.9, 1];
orientation: [−90, 90];
linear_err: [0, 100000];
area: [0, 900000];
c1:   Class contour;
c2:   Class contour;
c3:   Class contour;
c4:   Class contour;
c5:   Class contour.

Class SB7 upperclass SB
H: SB7;
T: [7, 7];
shape: spiral;
validity: good;
bar: present;
centring: indifferent;
arms: late.

Class E0   upperclass E
H: E0;
T: [−5,−5];
shape: elliptical;
bulge: visible;
validity: good;
profile_concavity: great;
flatness: null
bar: absent;
isophotes: smooth;
centring: good;
arms: absent.

The rules: There are 106 production rules. They represent the operating knowledge of the expert. The rules are used to affect a symbolic value to a descriptor of the object.

Example of rule:

Rule 8:

> **If** ellipticity of contour3 ≫ ellipticity 0.1
>
> ellipticity of contour3 > 0.4
>
> **then** bar is present

"As the ellipticity of the third contour is greater than the global ellipticity of the galaxy and is important, the bar is present".

There are several kinds of rules:

• rules used to interpret photometric parameters (such as *linear_err* or *profile*) and to affect a symbolic value to the parameters *bulge* or *profile_concavity*;

• rules used to interpret the shape of each contour according to the values of the parameters *ellipticity* and *compactness*;

• rules used to interpret the global shape of the galaxy according to the shape of each contour and the parameter *centre_err* of each contour;

• rules used to detect the presence of a bar according to the parameters *orientation* and *ellipticity* of the galaxy and *angle* and *ellipticity* of the contours;

• rules used to indicate the quality of the input parameters and thus of the classification according to the parameter *area*.

9.4.5.3 The fact base. The facts correspond to data relating to the current case and are temporarily added to the fact base; for our application, the facts represent information associated with one galaxy. The fact base is implemented with one object which has the same structure as the root prototype. Initially, all the values of its descriptors are unknown, except for those corresponding to the measured parameters. At the end of the procedure, the symbolic descriptors of the object have the same values as the prototype representing the class of the object.

9.4.6 Example of Session

Figure 9.9 shows the path in the hierarchy of the classes during the classification of NGC4473. The initial description of NGC4473 (measured parameters) is:

$$
\begin{array}{ll}
\text{orientation :} & -3.18 \\
\text{ellipticity:} & 0.4 \\
\text{linear\_err:} & 0.061 \\
\text{profile:} & 0.35 \\
\text{area:} & 10935 \\
\end{array}
$$

contours	centre_err	ellipse_err	compactness	angle	ellipticity
c1 :	0	0.07	1.6	−1.8	0.41
c2 :	1	0.04	1.9	−1.5	0.35
c3 :	0	0.05	3.2	−2.3	0.41
c4 :	0	0.05	3.8	−1.4	0.43
c5 :	5	0.09	6	−3	0.42

Twenty-four production rules (from the total of 106 rules) were activated in order to complete the description of the measured galaxy; the final symbolic description of the galaxy is:

Galaxy NGC4473:

class:	E4 (compatibility: total, incompatibility: null)
validity:	good (certain)
arms:	absent (certain)
shape:	elliptical (certain)
bulge:	visible (certain)
flatness:	faint (certain)
profile_concavity:	important (certain)
centring:	good (certain)
bar:	absent (certain)
isophotes:	smooth (certain)

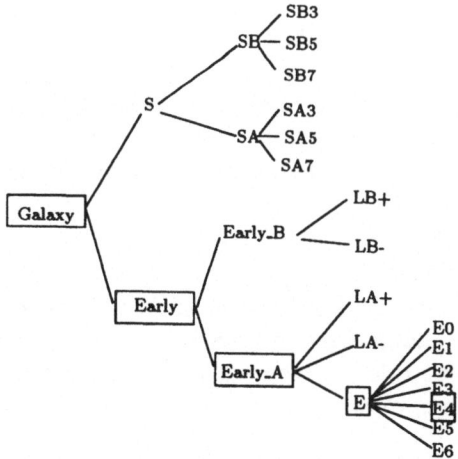

Fig. 9.9. The path in the tree traversed during the reasoning.

9.5 Results

A knowledge base of 37 prototypes and 106 rules has been built. The system has been tested on a set of 21 galaxies in the Virgo cluster digitized from the same Schmidt plate. The results of the automatic classification of these galaxies are presented in the following table:

Name	Actual class	Result	Comment
NGC4459	LAR+	LA+	correct
NGC4569	SXT2	SX3	correct
NGC4474	L...P*	LA+/LB+	correct
NGC4473	E.5..	E4	correct
NGC4477	LBS.*$	L	correct
NGC4571	SAR7	SA7	correct
NGC4468	LA...*	bad validity	small area
NGC4531	S..1	LA+	slight difference
NGC4419	SBS0	SB1/SA1	almost correct
IC3392	SA.3	SB1	almost correct (dust lanes)
M91	SBT3	SB1/SB3	correct
NGC4461	LBS+*	LA/LB	almost correct
NGC4438	SAS0	LB/SB1	correct (dust lanes)
NGC4421	SBS0	LB/SB1	correct
IC800	SBT4P$	SB5	correct
NGC4523	SBS8	SB	irregular
NGC4595	SXT3$	SA1/SA3	correct
NGC4639	SXT4	SB3	correct
NGC4540	SXT6	SB7	correct
M86	.E.3..	E3	correct
M84	.E.1..	E0	correct

Although the results are quite good, there remains the problem of false detection of a bar in the presence of transversal dust lanes.

Tests have shown that the method needs a minimal size for the galaxy image (about 50 × 50). So, the rule:

if area < 2500
then validity bad

is activated to prevent misclassification; in order to decrease the computing time, the parameter *validity* is set to *good* for each successor node of the root node *Galaxy*; so, if this rule is activated, no successor may be validated and the classification returns *class unknown* irrespective of the values of the other measured parameters.

The results obtained on a set of galaxies observed with another telescope (Schmidt telescope at Haute Provence Observatory) and with different exposure times show that the knowledge base must contain information about these symbolic parameters (*telescope* and *exposure_time*). As a result of the values of these parameters, different branches of the tree of prototypes must be scanned.

In its current state, the system classifies a galaxy in approximately 122 seconds: 120 seconds for the extraction of parameters (which is of course highly dependent on the size of the image) and 2 seconds for the inference phase.

9.5.1 Limitations

SYGAL is a completely automatic method for classifying complex objects (galaxies), using an expert system approach. This system, which has a knowledge base composed of 37 prototypes and 106 rules, should be extended to enhance the precision of classification but the classes currently obtained by this system are quite good.

Yet SYGAL, at this stage of its development, still has strong limitations before it can be considered to be a representative of a real galaxy classification system. Several kinds of limitations can be noted:

Artificial intelligence tool: The artificial intelligence tool used was a first version of CLASSIC, for which there was no possibility to go back to a low level during the classification reasoning in order to run specialized procedures.

In SYGAL, image processing was not supervised by an intelligent system specialized in monitoring (like an expert system built with OCAPI).

Algorithms: There are no peak or ridge detection algorithms. There are no real arm detection techniques with the exception of analysis of contour shape; so, there is no arm/interarm measure. There are no ring pattern detection or extraction algorithms; for similar reasons, there is not a real bulge/disk ratio measure with the exception of a measure of the degree of visibility of the bulge on the projected profile and the analysis of the contour shape.

Knowledge base: The knowledge base does not yet contain information about irregular and peculiar galaxies and about ring pattern structures. This lack of information is mainly due to the absence of such galaxies in the sampling set, so some further work is needed in this domain. Because of the resolution of the images used for the development of SYGAL, there was no knowledge of arm/interarm ratio and of the structure of arms.

9.6 Validation

A real automated system for the classification of galaxies needs to be strictly validated. We can distinguish four kinds of validation:

Algorithm validation: Validation, here, concerns expertise in vision and image processing. Generality, applicability, robustness, precision of results, and meaning of results of each procedure have to be checked.

Extracted parameter interpretation validation: The output of the image processing algorithms are used in the production rules or even in the prototypes; the validation of the knowledge which performs the passage of these measured parameters to the astronomical domain implies expertise in both vision (for good interpretation of the output of the algorithms) and in astronomy (for understanding of the astronomical features).

Classification knowledge validation: The classification knowledge expressed in the tree of prototypes, and in the prototypes themselves, is astronomical knowledge; thus this part needs to be validated by astronomers specialized in galaxy classification. This validation has to check the completeness, the correctness and the coherency of the expertise.

Image validation: The validation of an expert system (even if each part of the knowledge is checked by the experts with relevant skills) is carried out according to one or several sets of data (images). So, good validation needs large sets of homogeneous data with their context (observational conditions, image formation conditions, etc.), their class and their calibration data. This particular skill relates to observational expertise.

9.7 Conclusion

We have seen that the classification of galaxies is a very complex task but that it is not impossible to automate it. This can be useful both to objectively process large sets of data and to centralize and structure different kinds of expertise (in image processing, observation and classification); this last point becomes crucial with the multiplication of sensors used and the consequent difficulty of performing multisensor analysis.

The development of instrumentation (an ever increasing number of galaxies are studied with CCDs) will give us much new information. Knowledge of the morphology of these objects will greatly increase. This will surely lead us to refine prevalent classification schemes.

The method described in this chapter can easily be extended to the study of objects other than galaxies; for instance to the study of clusters.

References

1. Accomazzi, A., Bordogna, G., Mussio P. and Rampini A. (1989), "Rule based description and plausible classification of objects in digitized images", *Data Analysis in Astronomy III*, V. Di Gesù, L. Scarsi, P. Crane, J.H. Friedman, S. Levialdi and M.C. Maccarone (eds.), Plenum Press, New York, in press.
2. Bijaoui, A. (1980), "Sky background estimation and application", *Astronomy and Astrophysics*, **84**, 81–84.
3. Bijaoui, A. (1988) "Sampling effects in the analysis of astronomical objects", *IMACS 12th World Congress on Scientific Computation*, R. Vichnevetsky, P. Borne, J. Vignes (eds.), Vol. II, 316–317
4. Binggeli, B., Sandage, A. and Tamman, G.A. (1988), "The luminosity function of galaxies", *Annual Review of Astronomy and Astrophysics*, **26**, 509–650.
5. Buta, R. (1988), "Galaxy morphology and classification", *Le Monde des Galaxies*, Springer-Verlag, Heidelberg.

6. Cipiere, P. (1984), "Manuel de reférence du logiciel INRIMAGE", *INRIA*.

7. Debray, B. (1988), "Photométrie stellaire dans les champs encombrés pour l'étude des galaxies proches", Thesis, University of Aix-Marseille III.

8. Di Gesù, V. and Maccarone, M.C. (1986), "Description of fuzzy images by convex hull technique", *Proceedings of the 8th International Conference on Pattern Recognition ICPR9, Paris, France*, Vol. 2, IEEE Computer Society Press, New York, 1276–1278.

9. Dressler, A. (1980), "Galaxy morphology in rich clusters", *Astrophysical Journal*, **236**, 351–365.

10. Faber, S. (1987) *Dark Matter in the Universe*, J. Kormendy (ed.), G.K. Knapp, 1–16.

11. Fairfield, J. (1989), "A nearly invisible grid", *Computer Vision and Pattern Recognition*, submitted.

12. Freeman, K.C. (1975), *Galaxies and The Universe*, A. Sandage, M. Sandage and J. Kristian (eds.), University of Chicago Press, Chicago, 409–508.

13. Granger, C. and de Mongareuil, A. (1987) "The reference manual", *ILOG*.

14. Grosbøl, P. (1987), "Extraction of morphological parameters from images of galaxies", *Astronomy from Large Databases: Scientific Objectives and Methodological Approaches*, F. Murtagh and A. Heck (eds.), European Southern Observatory, Garching bei München, 107–110.

15. Hanson, A. and Riseman, E. (1978), "VISIONS: A Computer System for Interpreting Scenes", *Computer Vision Systems*, A. Hanson and E. Riseman (eds.), Academic Press, New York, 303.

16. Hayes-Roth, F., Waterman, D.A. and Lenat, D.B. (1983), *Building Expert Systems*, Addison-Wesley, Reading, MA.

17. Holmberg, E. (1975), *Galaxies and The Universe* A. Sandage, M. Sandage and J. Kristian (eds.), University of Chicago Press, Chicago, 123–158.

18. Hubble, E.P. (1926), *Astrophysical Journal*, **64**, 321.

19. Hubble, E.P. (1936), *The Realm of the Nebulae*, Yale University Press, New Haven.

20. Jarvis, J.F. and Tyson, J.A. (1981), "Focas: Faint Object Classification and Analysis System", *Astronomical Journal*, **86**, 476

21. Johnson, H.M. (1961), "Photographic photometry of S0 galaxies", *Astrophysical Journal*, **133**, 314–321.

22. Jones, W.B., Obbits, D.L., Gallet, R.M. and de Vaucouleurs, G. (1967), *Publications of the Department of Astronomy*, University of Texas, Austin Series II 1 no. 8.

23. Kormendy, J. (1980), *Two Dimensional Photometry*, P. Crane and Kjär (eds.), European Southern Observatory, Garching bei München, 191–238.

24. Kurtz, M.J., Huchra, J.P., Beers, T.C., Geller, M.J., Gioia Maccaro, T., Schild, R.E. and Stauffer, J.R. (1985), "The X-ray cluster Abell 744", *Astrophysical Journal*, **90**, 1665–1675.

25. Le Fèvre, O., Bijaoui, A., Mathez, G., Picat, J.P. and Lelièvre, G. (1986), "Electronographic BV photometry of three distant clusters of galaxies", *Astronomy and Astrophysics*, **154**, 92–99.

26. Lin, C.C., Yuan, C. and Shu F. (1969), "On the spiral structure of disk galaxies", *Astrophysical Journal*, **155**, 721–746.

27. MacGillivray, H.T. and Stobie, R.S. (1984), "New results with the COSMOS machine", *Vistas in Astronomy*, **27**, 433–475.

28. Martinet, L. (1985), "Contemporary dynamical problems — a possible contribution to their solution by galaxy photometry", *New Aspects of Galaxy Photometry*, J.L. Nieto (ed.), Springer-Verlag, Berlin, p. 121–130.

29. Oort, J. (1984), *The Big Bang and Georges Lemaitre*, A. Berger (ed.), D. Reidel, Dordrecht, p. 299.

30. Peebles, P.J.E. (1980), *The Large Scale Structure of the Universe*, Princeton University Press, Princeton, New Jersey.

31. Pouget, S., Thonnat, M., Clément, V. and de Fombelle, A. J. (1988), "Automatic diagnosis of a technical device by classification, the expert system DANTE: an application to antenna failures", *6th International Conference on Reliability and Maintainability*, Strasbourg, France.

32. Rocca-Volmerange, B. (1985), "Far-UV to infrared photometric star formation traces", *New Aspects of Galaxy Photometry*, J.L. Nieto (ed.), Springer-Verlag, Berlin, 199–210.

33. Rosenfeld, A. (1969), *Picture Processing by Computer*, MacGraw-Hill, New York, p. 127.

34. Sandage, A. (1961), *Hubble Atlas of Galaxies*, Carnegie Institution of Washington, Pub. 618, Washington, D.C.

35. Sandage, A. (1969), "The reddening, age difference, and helium abundance of the globular clusters M3, M13, M15, and M92", *Astrophysical Journal*, **157**, 515–531.

36. Sandage, A. (1975), *Galaxies and the Universe*, A. Sandage and J. Kristian (eds.), University of Chicago Press, Chicago.

37. Sersic, J.L. (1982), *Extragalactic Astronomy*, D. Reidel, Dordrecht.

38. Shortliffe, E.H. (1976), *Computer-Based Medical Consultation: MYCIN*, American Elsevier, New York.

39. Silk, J. (1986), *Sky and Telescope*, **72**, 582–587.

40. Simien, F. and de Vaucouleurs, G. (1986), "Systematics of bulge-to-disk ratio", *Astrophysical Journal*, **302**, 564–578.

41. Slezak, E., Bijaoui, A. and Mars G. (1988), "Galaxy counts in the Coma supercluster field", *Astronomy and Astrophysics*, **201**, 9–20.

42. Titterington, D.M. (1985), "General structure of regularization procedures in image reconstruction", *Astronomy and Astrophysics*, **144**, 381–387.

43. Thonnat, M. and Llebaria, A. (1981), *Astronomical Photography*, J.L. Heudier and M. Sim (eds.), CNRS-INAG Paris, 251–260.

44. Thonnat, M. (1982), *Traitement Numérique des Images Astronomiques*, A. Bijaoui (ed.), 255–279.

45. Thonnat, M. and Llebaria, A. (1983), *Applications of Digital Image Processing*, SPIE Proceedings Vol. 397, 454–461.

46. Thonnat, M. (1985), "Automatic morphological description of galaxies and classification by an expert system", *Rapport de Recherche*, INRIA No. 387.

47. Thonnat, M. (1989), "Toward an automatic classification of galaxies", *Le Monde des Galaxies*, Springer-Verlag, Berlin, in press.

48. Thonnat, M. and Clément, V. (1988), "OCAPI: a monitoring tool for the automatic control of image processing procedures" *Proceedings of the 12th IMACS World Congress on Scientific Computation*, Vol. II, R. Vichnevetsky, P. Borne and J. Vignes (eds.), 318–321.

49. Thonnat, M. and Clément, V. (1989), "OCAPI: an artificial tool for the automatic selection and control of image processing procedures", *Data Analysis in Astronomy III*, V. Di Gesù, L. Scarsi, P. Crane, J.H. Friedman, S. Levialdi and M.C. Maccarone (eds.), Plenum Press, New York, in press.

50. Thonnat, M. and Gandelin, M.H. (1988), "An expert system for the automatic classification and description of zooplanktons from monocular images", *Proceedings of the 9th International Conference on Pattern Recognition ICPR9*, Rome, Italy, November 1988.

51. de Vaucouleurs, G. (1948), *Annales d'Astrophysique*, **11**, p. 247.

52. de Vaucouleurs, G. (1959), "Classification and morphology of external galaxies", *Handbuch der Physik*, **53**, p. 275.

53. de Vaucouleurs, G. (1970), *Science*, **167**, 1203–1213.

54. de Vaucouleurs, G. (1976), *Le Monde des Galaxies*, Observatoire de Besançon et Laboratoire d'Astronomie de la Faculté des Sciences.

55. de Vaucouleurs, G. (1976), de Vaucouleurs, A. and Corwin, H.G. Jr. *Second Reference Catalogue of Bright Galaxies* (RC2), Texas University Press, Austin.

56. Watanabe, M., Kodaira, K. and Okamura S. (1982), "Digital surface photometry of galaxies: toward a quantitative classification", *Astrophysical Journal Supplement Series*, **50**, 1–22.

57. Winston, P.H. (1987), *Artificial Intelligence*, Addison-Wesley, Reading, MA.

58. Wolf, M. (1908), *Publ. Ap. Inst. König., Heidelberg*, **3**, No. 5.

10 Classification of IUE Spectra: A Rule Based Approach

Fionn Murtagh* (1), André Heck (2) and Roberto Rampazzo (3)

(1) Space Telescope — European Coordinating Facility
European Southern Observatory
Karl-Schwarzschild-Str. 2
D-8046 Garching bei München
F.R. Germany

(2) Observatoire Astronomique
11, rue de l'Université
F-67000 Strasbourg
France

(3) Osservatorio Astronomico di Brera
Via Brera 28
I-20121 Milano
Italy

10.1 Introduction

The increasing availability of large sets of reasonably homogeneous data (originating in the same instrument configuration, using the same calibration procedure, and being accessible through the same database system) motivate the use, not only of computer-aided statistical and pattern recognition analysis techniques (cf. Murtagh and Heck, 1987, 1988), but additionally of newer methodologies which try to emulate the behaviour of human experts.

---

*Affiliated to the Astrophysics Division, Space Science Department, European Space Agency.

Important justifications for using rule-based expert systems for classification purposes relate to the user interface; to the close relation between rules and physical interpretation; and to the facility offered by rules for expressing an expert's knowledge. These three justifications will be briefly looked at in turn.

Tools for the design and customizing of a user interface are normally part of an expert system. The user interface can be modified as the system expands. It can cater well for the inexperienced user, given its use of graphical presentation of information and user interaction via a mouse, icons and so on. User interaction with the system is often more convenient compared to more traditional user interface paradigms (commands followed by parameters or simple multiple choice menus).

Rules in expert systems can also be meaningfully related to the physics involved in the problem under investigation. The latter can guide the formulation of rules and subsequently the classification arrived at can be related to the underlying physics through these rules.

Knowledge capture may be problematic in the context of many expert system designs. However, notwithstanding this, it is often easier to ask an astronomer for the "intuitive" rules used for, e.g., spectral classification. The alternative is often a (mathematical, statistical, or other) model which may hinder rather than help communication between knowledge engineer and domain expert.

Classification of celestial objects is one of the central themes in astronomy and artificial intelligence techniques have already demonstrated promise for this problem (Bernat and McGraw, 1986; Thonnat and Berthod, 1984). The problem of classification of stellar spectra in the ultraviolet originates in the fact that the Morgan-Keenan classification scheme (the MK classification system: see *inter alia*, Henden and Kaitchuk, 1982; Keenan, 1987), defined in the visual range, cannot be immediately extended to wavelength regions for which it is not defined. If the classification scheme itself cannot be directly used, the general approach employed may be. This by-now classical approach requires a simultaneous categorization into spectral type and luminosity class (Jaschek and Jaschek, 1984, 1987). Additionally, the increasing quality and quantity of data available force the astronomer to consider various computer-aided classification approaches (Kurtz, 1983; Heck et al., 1986a).

Low resolution ultraviolet stellar spectra obtained with the International Ultraviolet Explorer satellite (IUE, funded by ESA, NASA and the British SERC, which recently celebrated a decade of active life) can be selected in fairly homogeneous sets, although they do not have the characteristics of a survey. The latter would involve comprehensiveness and exhaustivity within predefined limits (of location, magnitude or object type). The ultraviolet classification of "normal" stars has already been defined (Heck et al., 1984) and that of "peculiar" stars has been started. The latter has not yet been completed due to the even larger lack of correspondence between the visual and the ultraviolet spectral ranges (Heck et al., 1986a; Heck et al., 1986b; Heck, 1987). As just one example of the importance of having a correct classification of ultraviolet stellar spectra, mention may be made of its importance for the determination of interstellar extinction (see, e.g., Carnochan, 1986, and references therein).

We present here the backbone of a *rule-based classifier*, the aim of which is to reproduce automatically the IUE classification, with as input wavelength calibrated spectra. In section 2 we discuss the input data used in the system as it currently operates, and in section 3 we describe the system structure.

10.2 Spectral Data Used

Tests of the classifier have, so far, been performed on spectra (5 Å resolution) from Heck et al. (1984). The initial system design was carried out on the basis of approximately 250 well-classified spectra and 29 parameters derived from them. Of these, 28 are weighted line intensities in the region from 1175 to 2855 Å (see Table 10.1); and one, the asymmetry coefficient, was a parameter which describes the continuum and on which a reddening correction had been carried out.

The parameters and their determination have been described in Murtagh and Heck (1987). This last reference used multivariate data analysis techniques (cluster analysis, principal components analysis and discriminant analysis) on the IUE spectra.

Table 10.1. Lines used for classification: wavelengths and associated elements (5 Å resolution)

1175 C III	1400 Si IV	1630 Fe IV	1925 Fe III
1215 H I	1430 Fe V;Si II	1640 O IV	1935
1255 Fe V	1455 Fe V	1720 C II; Al II	1965 Fe III
1265 Si II	1465 Fe V	1725 Fe IV	2670
1300 Si III	1550 C IV	1850	2755
1335 C II	1610 Fe II	1855 Al III	2800 Mg II
1370 Fe V;O V	1620 Fe V	1890 Fe III	2855 Mg II

The lines in the table are in most cases a blend of lines. In the system implemented, a help facility is available for such blends. Some of these lines may be of interstellar or geocoronal (such as L_α) origin, but not necessarily along the entire sequences of spectral types and luminosity classes.

The determination of the rules that govern the reasoning are derived from the same data as that used for multivariate data analyses (see references above). Low level rules, attempting to identify roughly the general category (early, intermediate, late-types) to which the spectrum may be assigned, are based on this continuum trend.

10.3 Prototype System Structure

The system sketched out in Fig. 10.1, now partly implemented, analyzes the spectrum by building an initial command language procedure (in the European Southern Observatory's MIDAS — Munich Image Data Analysis System — image processing system) which roughly measures the features of the spectrum (e.g. the "asymmetry coefficient"). A table of values is passed to the expert system which uses low level rules to obtain general-type information (e.g. that the spectrum under investigation is a late type star). From these rules the expert system gives information on the lines that have not been found, but potentially are present; it then builds a new command procedure to instruct the host system to find and measure these line intensities. A more precise classification is derived and a comparison with standard stars in the knowledge base gives an indication as to when to interrupt this iterative improvement.

The basic parts of the system structure are the *User Interface*, the *Explanation Subsystem* and the *Knowledge Base*. Each of these will be examined in turn.

User Interface: This is icon and mouse-oriented, using the KEE (*Knowledge Engineering Environment*) system produced by IntelliCorp. KEE is a hybrid object oriented tool. That is to say, it allows the use of programming languages such as Lisp in addition to the features provided by the system itself; and KEE's frames are used to define the basic objects — the building blocks such as "early-type-star" — on which the system is based. For general articles describing systems built with KEE see Faught (1986) and Fikes and Kehler (1985); or for comparative appraisals of expert system shells, see, e.g., Cross (1986) or Gilmore et al. (1986).

One among a number of panels, the explanation panel, relates to the relevant spectral lines using the explanation subsystem (see below) in order to state, for example, that:

> *The criteria used for the classification are:*
> • *the value of the asymmetry-coefficient corresponds to that of an early-type star,*
> • *the 1255 [Fe V] line is weak,*
> • *the 1620 blend is weak or moderately strong,*
> • *and so on.*

Knowledge Base: This includes standard stars and their properties for each class which the system uses; and the rules used for the classification. There are approximately 50 such rules at present of the form:

> **IF** *the asymmetry coefficient is in a particular range, a particular line is present, another line is in a particular range, etc.*
> **THEN** *the spectra is in a class specified by a spectral type and a luminosity class.*

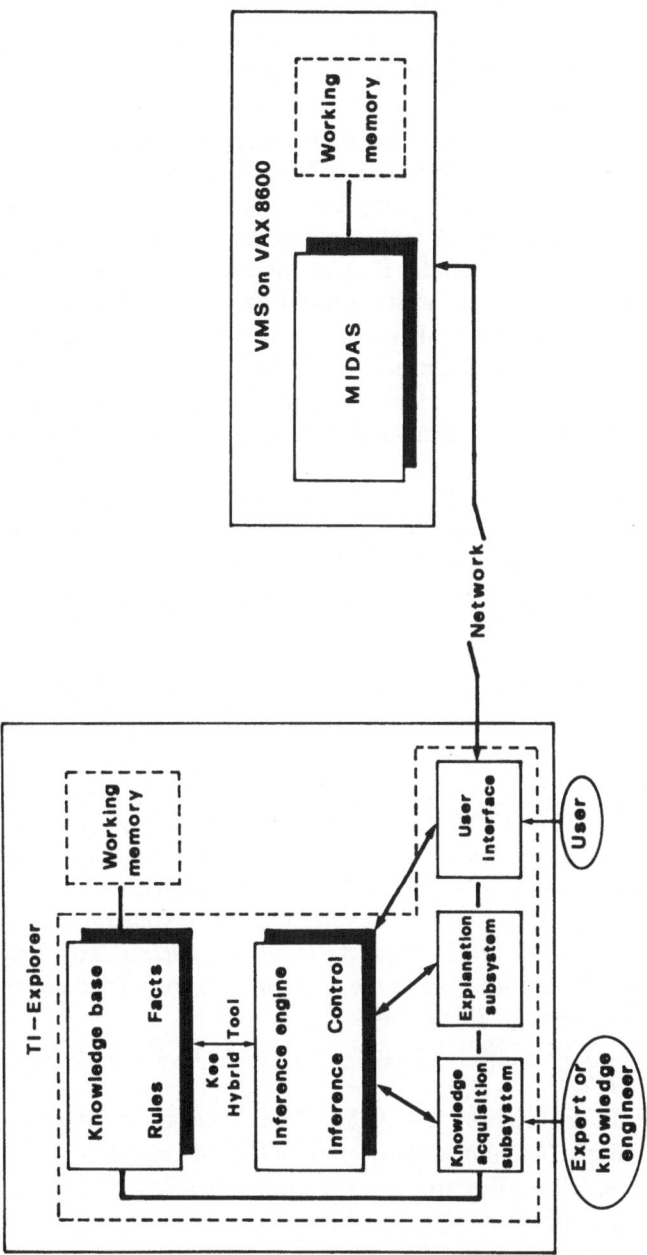

Fig. 10.1. Schematic structure of the expert system and interaction with the host. The philosophy is that all operations involving computations are performed using well established and tested procedures in the MIDAS command language, with only the "reasoning" and classification being demanded of the expert system. This latter, other than building batch procedures to analyse spectra, controls input/output and gives explanations to the user about the criteria used in the classification process.

Explanation Subsystem: This contains, for example, the spectral lines expected and associated information, irrespective of whether or not the lines are used in the classification procedure. There are approximately 600 such lines, described in the present system, for early type stars alone.

An initial hierarchy of rules which starts identifying a gross spectral classification, starting from the values of the asymmetry coefficient and achieving a detailed classification by working on the line intensities, has been implemented and tested on a set of 229 well classified spectra (Heck et al., 1984). The rules to classify a set of normal O, B, A, F, and G supergiant, giant and dwarf stars (the latter map onto luminosity classes) have been developed.

Along the lines previously sketched out, a basic but satisfactory user interface (selection panel, control panels, input/output with the MIDAS environment) and an explanation subsystem have been built. More flexible communication with the MIDAS image processing environment is necessary and is planned.

Acknowledgements. H.-M. Adorf provided invaluable assistance at all stages. Discussions with R. Albrecht and M. Johnston are appreciated. P. Benvenuti encouraged and supported this work. One of us (R.R.) gratefully acknowledges an ESA External Fellowship.

References

1. Bernat, A.P. and McGraw, J.T. (1986), "An intelligent object recognizer and classification system for astronomical use", *Instrumentation in Astronomy VI* (ed. D Crawford), SPIE Proceedings, **627**, 89–94.
2. Carnochan, D.J. (1986), "The variation of interstellar dust in the ultraviolet", *Monthly Notices of the Royal Astronomical Society*, **219**, 903–926.
3. Cross, G.R. (1986), "Tools for constructing knowledge-based systems", *Optical Engineering*, **25**, 436–444.
4. Faught, W.S. (1986), "Applications of AI in Engineering", *Computer*, **19**, July, 17–27.
5. Fikes, R. and Kehler, T. (1985), "The role of frame-based representation in reasoning", *Communications of the ACM*, **28**, 904–920.
6. Gilmore, J.F.A., Howard, C. and Pulaski, K. (1986), "A comprehensive evaluation of expert system tools", *Applications of Artificial Intelligence IV*, SPIE Proceedings **657**, 194–208.
7. Heck, A. (1987), "UV stellar spectral classification", *Exploring the Universe with the IUE Satellite* Y. Kondo et al. (eds.), D. Reidel Publishing Company, Dordrecht, 121–137.
8. Heck, A., Egret, D., Jaschek, M. and Jaschek, C. (1984), *IUE Low Resolution Spectra Reference Atlas, ESA SP-1052.*
9. Heck, A., Egret, D., Nobelis, Ph. and Turlot, J.C. (1986a), "Statistical confirmation of the UV stellar spectral classification system based on IUE

low-dispersion stellar spectra", *Astrophysics and Space Science*, **120**, 223–237.

10. Heck, A., Egret, D., Hassall, B.J.M., Jaschek, C., Jaschek, M. and Talavera, A. (1986b), "IUE low-dispersion spectra reference atlas — Volume II. Peculiar stars", *New Insights in Astrophysics — Eight Years of UV Astronomy with IUE*, ESA-SP 263 (ed. E. Rolfe), 661–664.

11. Henden, A.A. and Kaitchuk, R.H. (1982), *Astronomical Photometry*, Van Nostrand Reinhold, New York, ch. 2.

12. Jaschek, C. and Jaschek, M. (1984), "Classification of ultraviolet spectra", *The MK Process and Stellar Classification* (ed. R. F. Garrison), David Dunlap Observatory, Toronto, 290–304.

13. Jaschek, C. and Jaschek, M. (1987), *The Classification of Stars*, Cambridge University Press, Cambridge, UK.

14. Keenan, P.C. (1987), "Spectral types and their uses", *Publications of the Astronomical Society of the Pacific*, **99**, 713–723.

15. Kurtz, M.J. (1983), "Progress in automation techniques for MK classification", *The MK Process and Stellar Classification* (ed. R. F. Garrison), David Dunlap Observatory, Toronto, 136–152.

16. Murtagh, F. and Heck, A. (1987), *Multivariate Data Analysis*, Kluwer, Dordrecht.

17. Murtagh, F. and Heck, A. (1988) (eds.), *Astronomy from Large Databases: Scientific Objectives and Methodological Approaches*, European Southern Observatory, Garching bei München.

18. Thonnat, M. and Berthod, M. (1984), "Automatic classification of galaxies into morphological types", *Proceedings of the 7th International Conference on Pattern Recognition*, IEEE Computer Science Press, New York, 844–846.

Important Methods

11 Representation of Knowledge Using Fuzzy Set Theory

Vito Di Gesù

Dipartimento di Matematica ed Applicazioni
Università di Palermo
Via Archirafi 34
I-90123 Palermo
Italy

and

Istituto di Fisica Cosmica ed Applicazioni dell'Informatica, C.N.R.
Via Mariano Stabile 172
I-90139 Palermo
Italy

11.1 Introduction

Sciences such as physics, engineering and chemistry build exact models of phenomena, by starting from experimental data, and use these models to make predictions. A very crucial point of this process is, of course, the correct analysis and evaluation of the experimental data. The models are mainly based on algebra and probability theory.

Data analysts also use mathematics and probability. However *human experience* may play a fundamental role whenever inexactness and vagueness are properties of the data. Moreover *cloudy* quantities exist which are not describable in terms of probability distributions, such as the *beauty of some thing* or the *tallness of a man*. In all these cases, education, fashion and global knowledge may play a crucial role in making decisions.

The development of physics as opposed to that of mathematics is based on claims supported by many facts (previous experimental results and previous models), which must be "confirmed" by new experimental results. The main difference between mathematics and physics is that as soon as a mathematical theorem has been proved true, it will hold once and for all, while physics is only able to hold temporary truths.

This uncertainty is the beauty of physics, from which it follows that in physics the model of human reasoning is based on subjectivity. The efforts of the scientists is to convince, with subjective reasoning, the greater part of their claims.

Even probability arguments are subjective; as a matter of fact, if used to support a hypothesis, they may only be used to define a confidence level for accepting or rejecting it. A first question will arise: *Which is the best confidence level?*

The situation becomes dramatic whenever the probability density functions (pdf) are not known or it is impossible to compute them from the experimental data (low statistics or very complex pdf). In such situations of ignorance, indicators based on the non-parametric evaluation of the information contained in the data could be more useful in taking decision.

For example the inner analysis (structure, morphology) of faint and extended celestial objects could profit from a fuzzy set approach whenever it is hard to answer questions like: *What is the probability that the shape of the object* **X** *is a spiral with two arms and a kernel of size d?* Or: *Which are the most relevant parameters in order to represent possible clusters of radiopulsars?* (The number of parameters ~ 10, number of samples ~ 336).

The first question shows an example in which astronomers want to handle linguistic variables rather than statistical variables. The second question is related to a situation in which the number of data samples is very small in comparison with the dimension of the variable space.

The steady growth of experiments (mostly in astronomy and high energy particle physics) has introduced new complexity in the information that is considered in the analysis. The evaluation of the conclusions may become very difficult. For example several models could be considered and compared with the results of the analysis. Then a new question could arise: *Given an experimental result which is the best model associated with it?*

The last question may have only a subjective answer, even if many "facts" can support it. For example Galileo's gravitation theory was supported by many experiments (Galileo Galilei, 1632) and confirmed later by other theories; however Einstein's gravitation theory (Einstein, 1916) supplanted it by explaining new experimental evidence. The statement *"Einstein's gravitation theory is true"* is meaningless as also is the statement: *"Einstein's gravitation theory is true with probability* p*"*. This is another example in which we would like to express an uncertain concept, without using probability.

Scientists want to compare and correlate their data and results with data and results obtained from other experiments. Hence there exists the need for large knowledge bases containing models, references to existing experiments and bibliographic data. Hence there exists also the need for systems for data analysis to guide data analysts in the navigation through the knowledge base and in taking the best subjective decisions.

Artificial intelligence tools and methods must be applied to achieve such goals (Rosenthal, 1988) whenever a knowledge based system (KBS) oriented to data analysis is used. KBSs may be defined as machines that make inferences from internalized facts and rules. The facts and the rules are chunks of knowledge (methods) or statements (models) about the external world (experimental data). Our aim is to give an overview of current methodologies applied in order to represent facts and rules with uncertain and linguistic values.

L. Zadeh (1962) introduced the concept of *fuzzy set*, as an extension of the classic ones. In the theory of fuzzy sets, the characteristic function, χ_A, of a subset A of a universe X ranges in the interval [0,1] instead of in the set $\{0,1\}$.

Given an element $x \in X$, $\chi_A(x)$ is also termed the degree of belonging of x to the fuzzy set A. The universe X could represent the measured data and χ_A could be an evaluation of the data with respect to the hypothesis A.

Since fuzzy sets were introduced, their mathematical foundation has been widely developed. Measure theory has been applied in order to extend on them topological spaces (Katsaras, 1981). An extension of information theory has been studied (De Luca and Termini, 1972, Dimitrov, 1983). *Possibility theory* has been developed in connection with probability theory (Yager, 1980a). Fruitful applications have been made in the fields of the artificial intelligence (Negoita, 1985), of image analysis and processing (Dutta Majumder, 1980), of linguistic analysis (Yager, 1980b) and multivalent logic (Kandel, 1976; Zadeh, 1985).

In particular the application of fuzzy sets to image analysis seems to be promising, if the data are sparse and/or the probability model is not available or difficult to define (Di Gesù, 1986; 1988). Fuzzy sets could be a good basis for handling such problems. The combination of probability and possibility theory could be used in order to analyse data, with complementary information coming from both sides.

Section 11.2 is dedicated to the concept of KBS. In section 11.3 some basic definitions and properties relating to fuzzy sets are introduced. Section 11.4 explores some implementation techniques useful to represent vague knowledge in KBSs. Section 11.5 shows applications in the analysis of *sparse images*. Concluding remarks are made in section 11.5.

11.2 KBS and Data Analysis

Several books and papers have been written to develop and to define the concept of KBS (Waterman, 1986). Some authors put the emphasis on the cybernetic aspects (thinking machines: Stefik, 1986; Michie, 1983; LaChat, 1986), others on applications (pragmatic approach: Thonnat and Clément, 1989; Shortliffe, 1976; Davis, 1980; and Lenat, 1982).

We use the term KBS to refer to a computer system, which incorporates one or more techniques of artificial intelligence to perform a family of activities that traditionally would have to be performed by a human expert. In Figure 11.1 a typical structure of a KBS is shown.

The *knowledge base* contains a collection of related facts, procedures, models and heuristics that can be used in problem solving or inference systems. Knowledge is then a refined kind of information, which is organized so that it can be readily applied to solving problems. It can be deterministic (exact, certain) or non-deterministic (uncertain information). The latter case is the most common and interesting in the analysis of experimental data. The elements of a knowledge base may be represented in several ways: relational database, frames, semantic nets, production rules (McCalla and Cercone, 1983); the choice of the organization depends on the nature of the problem and the cardinality of the data. A relational database (Ullman, 1982; 1976) allows the handling of n-ary relations and provides inference mechanisms (selection, projection, join). Frames and semantic networks (Woods, 1975) allow the handling of binary or ternary relations. These do not provide inference mechanisms and they require strong organization. Production rules are based on the relations *if...then*, they provide inference mechanisms (rule activation) and do not need strong organization.

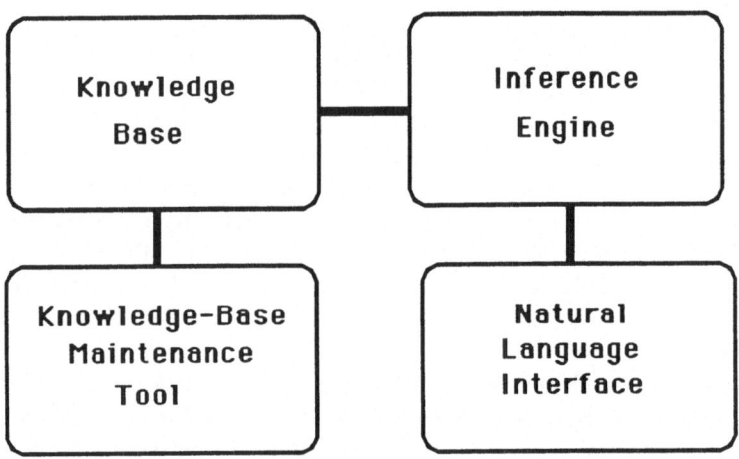

Fig. 11.1. The typical structure of a KBS.

The representation of uncertainty in the knowledge may be expressed in several ways depending on the choice of organization. If frames are used, uncertainty is represented in the attribute-value list, as probability measures or membership values.

The *inference engine* is the procedure which generates consequences, conclusions or decisions from existing knowledge. It may operate as a rule interpreter or as a routine which explores and updates the relations (arcs) in a non-deterministic inference network or tree. In the first operational mode the knowledge and the decisions are deterministic; while in the second, both knowledge and inference are uncertain. Such uncertainty may be modelled by probability (Bayesian inference rules) or by some measure of this uncertainty (fuzzy inference rules). The choice of one of the two models of uncertainty depends by the nature of the decision and the knowledge. The two models may be combined or alternatively used at different levels of the inference network (see Figure 11.2).

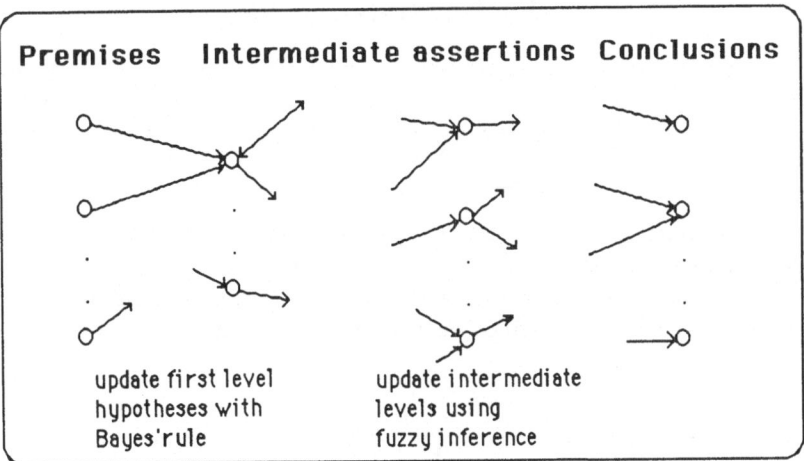

Fig. 11.2. Combined inference network.

KBSs, oriented to the solution of data analysis problems, interact often with expert users (consultant systems) due to the fact that fully automatic systems are not effective whenever decisions depend strongly on uncertain knowledge modelled by probabilistic or fuzzy laws. This is the reason why special attention must be paid to the design of natural language and advanced graphics interfaces. The latter requires the capability of internalization of linguistic variables and concepts. Fuzzy sets could provide new tools to represent and handle such kinds of knowledge.

11.3 Fuzzy Sets: Definitions and Properties

This section may be considered as a short tutorial on fuzzy set theory. All concepts introduced will be related to the problem of designing a KBS dedicated to data analysis. Readers familiar with the topic may skip it.

11.3.1 Algebra of Fuzzy Sets

Fuzzy sets may be formally defined in the framework of measure theory.

DEF1. Given a subset $\mathbf{A}$ of a universe $\mathbf{X}$ and a function $\chi : \mathbf{X} \to [0,1]$ then the set of ordered pairs $(x, \chi_A(x))$ is called a fuzzy set of $\mathbf{X}$, in symbols:

$$\mathbf{A} = \{x \mid (x, \chi_A(x)), x \in \mathbf{X}\}$$

It follows that ordinary sets may be considered as an example of fuzzy sets. $\chi_A(x)$ is termed the grade of membership, or degree of belonging, of x to $\mathbf{A}$. The elements with degree 1 are called the *ideal members* of $\mathbf{A}$, those with degree 0 are the *non-members* of $\mathbf{A}$.

It is sometimes useful to represent a fuzzy set as follows:

$$\mathbf{A} = \{x \mid x/\chi_A(x), x \in \mathbf{X}\}$$

The universe ($\mathbf{X}$) and the empty set ($\emptyset$) are defined as the ordinary ones as follows:

$$\mathbf{X} \iff \chi_\mathbf{X}(x) = 1 \quad \forall x \in \mathbf{X}$$

$$\emptyset \iff \chi_\phi(x) = 0 \quad \forall x \in \mathbf{X}$$

The interpretation and the values taken by χ depend on the a priori knowledge of the system being studied (physical model, subjective thinking) and on the goals of the analysis (data transformations, classification, presentation). Figure 11.3 shows two fuzzy sets defined on the real straight line.

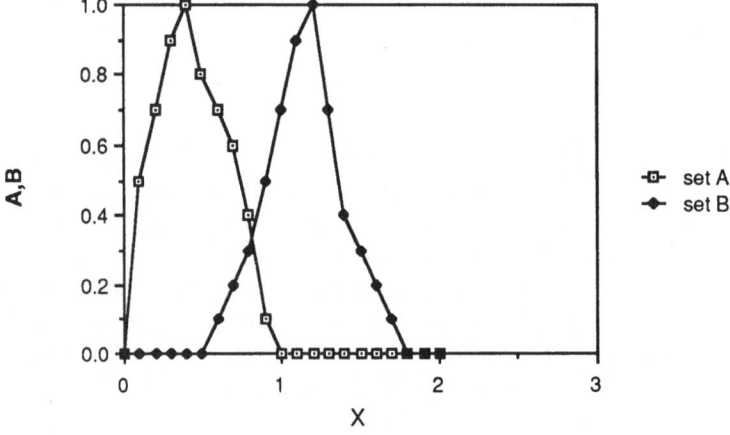

Fig. 11.3. Examples of fuzzy sets on the straight line.

Another example is related to the representation of fuzzy numbers. The fuzzy number 5 could be for example the fuzzy set:

$$\{0/0, 1/.1, 2/.5, 3/.6, 4/.7, 5/1, 6/.7, 7/.6, 8/.5, 9/.1, 10/0\}$$

An order relation may be established between fuzzy sets:

DEF2. Given two fuzzy sets $\mathbf{A}$ and $\mathbf{B}$ on the universe $\mathbf{X}$:

$$\mathbf{A} \subseteq \mathbf{B} \iff \chi_\mathbf{A}(x) \leq \chi_\mathbf{B}(x) \quad \forall x \in \mathbf{X}$$

Set operations may also be defined on fuzzy sets. Let $\mathbf{A}$ and $\mathbf{B}$ be two fuzzy sets, with membership functions $\chi_\mathbf{A}$ and $\chi_\mathbf{B}$ respectively. Then:

DEF3. The union, **C**, of **A** and **B** has the membership function:

$$\chi_{\mathbf{A} \cup \mathbf{B}}(x) = \max\{\chi_{\mathbf{A}}(x), \chi_{\mathbf{B}}(x)\}$$

DEF4. The intersection, **C**, of **A** and **B** has membership function:

$$\chi_{\mathbf{A} \cap \mathbf{B}}(x) = \min\{\chi_{\mathbf{A}}(x), \chi_{\mathbf{B}}(x)\}$$

DEF5. The complement, ¬**A**, of **A** has membership function:

$$\chi_{\neg\mathbf{A}}(x) = 1 - \chi_{\mathbf{A}}(x)$$

DEF6. The difference, **C**, of **A** and **B** has membership function:

$$\chi_{\mathbf{A}-\mathbf{B}}(x) = \min\{\chi_{\mathbf{A}}(x), 1 - \chi_{\mathbf{B}}(x)\}$$

It is easy to see that the union and the intersection are associative, commutative and mutually distributive. The complement is involutive (**A** = ¬¬**A**).

All these operations are equivalent to the classic ones if applied to an ordinary characteristic function. However, on the basis of its definition, it must be noted that:

$$\mathbf{A} \cap \neg\mathbf{A} \neq \emptyset \quad \text{(empty set)}$$
$$\mathbf{A} \cup \neg\mathbf{A} \neq \mathbf{X} \quad \text{(universe).}$$

These inequalities may express, for example, the fact that a single model and its negation do not exhaust all possible models of a physical problem.

Figure 11.4 shows the union, intersection and complement of the fuzzy set shown in Figure 11.3.

Authors have introduced other sets of operations, useful in the design of decision systems, such as product and exponentiation (Kaufmann, 1980).

The reader may find a complete exposition of the algebra of fuzzy sets in Kandel's book (Kandel, 1986).

The algebra of fuzzy sets has been applied recently to extend mathematical morphology (Serra, 1982) to multigray level digital images (Di Gesù, 1989). The main idea is to consider an image, after opportune normalization of the pixel values (Dutta Majumder, 1986), as fuzzy sets. From this point of view, all the operations previously introduced could be interpreted as transformations on images.

The algebra of fuzzy sets allows us to define classifiers with mixed classes defined as follows: given a universe **X** and $1, 2, \ldots, m$ fuzzy sets with the ideal element, then $\forall x \in \mathbf{X}$ a mixed classification of x is the fuzzy set $\{\chi_1(x), \chi_2(x), \ldots, \chi_m(x)\}$ built on the integer universe $\{1, 2, \ldots, m\}$. Mixed classification may be useful whenever a complete separation between classes does not exist, or for the search for relevant features in the **X** space. In this case the search algorithm is based on the minimization of the mean variance of the classifier defined as:

$$\overline{\mathbf{Var}} = \frac{1}{|\mathbf{X}|} \sum_{x \in \mathbf{X}} \sum_{i=1}^{m} [\chi_i(x) - \overline{\chi(x)}]^2.$$

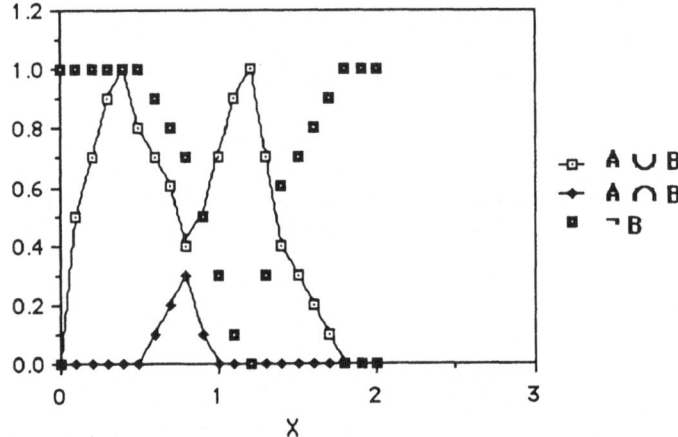

Fig. 11.4. Examples of fuzzy operators.

The definition of fuzzy graph has been also introduced (Rosenfeld et al., 1976):

DEF6. Given a graph $\mathbf{G} =< \mathbf{V}, \mathbf{E} >$, where $\mathbf{V}$ is the set of nodes and $\mathbf{E}$ is the set of arcs, a *fuzzy graph*, $\mathcal{G}$, is a pair $< \mathcal{V}, \mathcal{E} >$, where $\mathcal{V}$ is a fuzzy set on $\mathbf{V}$ and $\mathcal{E}$ is a fuzzy relation on $\mathbf{V} \times \mathbf{V}$ such that:

$$\chi_{\mathcal{E}}(x, y) \leq \min\{\chi_{\mathcal{V}}(x), \chi_{\mathcal{V}}(y)\}.$$

Fuzzy graphs are used in the solution of *fuzzy clustering* problems, where the degree of connectivity between two nodes, x and y, is determined by computing the length of the shortest path, $x \equiv x_0, x_1, \ldots, x_k \equiv y$, between them, defined as the maximum of the $\chi_{\mathcal{E}}(x_i, x_{i+1})$ for $0 \leq i < k$.

The concept of connectivity is also used in the framework of the design of a KBS in order to formalize subjective decision models. In this case the nodes may be considered as facts or rules, and their membership functions represent their consistency with the system model. For example in the analysis of spectra the facts could be statements about a set of spectra themselves, such as the existence of lines (energy and intensity) or the kind of spectrum (power, exponential, etc.). Probabilities are associated with each line and errors are associated with the fitting of each spectrum model. Such quantities may be interpreted as membership functions of facts with the data. Other facts could be theoretical models of star emissions. In this case the membership function of each fact represents the subjective plausibility of each emission model. In a situation of complete ignorance its value may be put equal to $0 < \alpha \leq 1$, for all emission models. In this example the *fuzzy relation* expresses the following reasoning:

Given a spectrum, S, with properties $\{s_1, s_2, \ldots, s_k\}$ and membership function α, given the emission model M with membership function β, then the plausibility of the statement: "M agrees with S" is $\gamma = \min\{\alpha, \beta\}$.

The fuzzy graph may be visited to evaluate the production rules of a production system. The inference may be carried out by using the same $\max - \min$ rule used to solve the clustering problem:

condition on $\max\{\min\{\alpha_i, \beta_j\}\} \Rightarrow$ **decision**.

The computational complexity of the evaluation procedure may be reduced if, as a preliminary, all arcs satisfying $\gamma = \min\{\alpha, \beta\} < \emptyset$ are deleted.

11.3.2 Fuzzy Logic and KBS

The evaluation of conditions can also be carried out by using *fuzzy logic* (Dubois and Pradé, 1979a; Rescher, 1968), by assuming the truth value of a proposition $P, \mathbf{v}(P) \in [0, 1]$. It follows that fuzzy logic is also multivalent. Three main types of logic have been widely studied in the literature (Kandel, 1986). One of the most natural is based on the following definitions:

$$\mathbf{v}(P \vee Q) = \max(\mathbf{v}(P), \mathbf{v}(Q)),$$

$$\mathbf{v}(P \wedge Q) = \min(\mathbf{v}(P), \mathbf{v}(Q)),$$

$$\mathbf{v}(\neg P) = 1 - \mathbf{v}(P)$$

It is clear that $\vee$ and $\wedge$ are commutative, associative, idempotent, and distributive over one another, and do not satisfy the excluded-middle laws in the sense that $\mathbf{v}(P \vee \neg P) \neq 1$, $\mathbf{v}(P \wedge \neg P) \neq 0$; moreover the absorption, De Morgan, equivalence and exclusive disjunction laws hold. Hence fuzzy logic is quite powerful and allows us to express complex uncertainty conditions. Valuation for quantifiers are straightforwardly defined as:

$$\mathbf{v}(\forall x P(x)) = \inf(\mathbf{v}(P(x))$$

$$\mathbf{v}(\exists x P(x)) = \sup(\mathbf{v}(P(x))$$

where x denotes an element of the universe of discourse.

Further developments of fuzzy logic allow approximate reasoning and linguistic approximation to be formalized. Both of these concepts are widely applied for emulating human thinking and expressing queries and retrievals in simple natural language. The key to this is the definition of *semantic equivalence*.

Let P and Q be two propositions and let Π_P and Π_Q be the truth distributions (also called possibility distribution) induced by P and Q. Then P and Q are said to be semantically equivalent if $\Pi_P = \Pi_Q$, which is denoted by $P \Longrightarrow Q$. *Semantic implication*, $P \Rightarrow Q$, holds iff $\Pi_P \subseteq \Pi_Q$. Here, of course, set inclusion and equality are defined in the framework of fuzzy set theory.

Natural languages include quantifiers, which express at several levels the extension of a given proposition. They may be considered in between $\forall$ and $\exists$. Examples are: *several, most, much, many, not many, not very many, few*. Such quantifiers may be represented as fuzzy numbers and manipulated by using fuzzy algebra so that:

$$\chi_{\text{not-many}} = 1 - \chi_{\text{many}}.$$

For example if *many* is represented by the fuzzy number

$$\{10/.1, 20/.3, 30/.6, 40/.8, 100/1\}$$

then *not-many* is represented by

$$\{10/.9, 20/.7, 30/.4, 40/.2, 100/0\}.$$

The attribute *very* may be expressed by squaring the value of the fuzzy quantifier, to which it is applied. Following the previous example *not-very-many* is the fuzzy number $\{10/.81, 20/.49, 30/.09, 40/.04, 100/0\}$. The last two examples show the power of fuzzy sets for modelling subjectivity.

Another example of the use of fuzzy logic concerns the representation of linguistic quantities. Suppose that a researcher wants to express the fact that the values of a measured quantity, W, give some indication about a physical effect (but not the evidence), and suppose that he is able to assign to each measure, w, an *a priori* degree of evidence, $\chi_E(w)$ (hypothesis H_0) and an *a priori* degree of normality, $\chi_N(w)$ (hypothesis H_1). We can now define the following linguistic variables as fuzzy sets:

$$evidence = \{w \mid w/\chi_E \quad w \in \mathbf{W}\}$$

$$normality = \{w \mid w/\chi_N \quad w \in \mathbf{W}\}$$

and by new linguistic variables built from simple fuzzy logic operators:

$$non - evidence = \{w \mid 1 - w/\chi_E \quad w \in \mathbf{W}\}$$

$$non - normality = \{w \mid 1 - w/\chi_N \quad w \in \mathbf{W}\}.$$

Finally:

$$some - evidence$$

$$= non - evidence \text{ and } non - normality$$

$$= \{w \mid \min(1 - w/\chi_N, 1 - w/\chi_N) \quad w \in \mathbf{W}\}.$$

A KBS with production rules based on fuzzy logic should now be able to answer a query about the quality of some measured value, w, by the evaluation of the following IF-THEN rule:

if $\chi_M(w)$ *is almost equal to some-evidence*
then *"some evidence of a non-normal signal exists in the measured data"*.

11.3.3 Rules of Inference

A principal conclusion of the previous section on knowledge representation is that before a KBS can deal with one sentence, all of it must be translated into an internal description by using fuzzy numbers or variables with degrees of truth that capture its meaning (semantic value).

After this translation, the KBSs must be able to carry out inference by starting from some premises. In the framework of classic logic the resolution rule is often applied. By using the *modus ponens* rule, it may be stated as follows:

$$P \to Q \quad and \quad P \vdash Q$$

which means that if $(P \to Q)$ is true and if P is true then we can conclude that Q is true. An extension of this is the *chain rule*:

$$(P \to Q \quad and \quad Q \to R) \quad \vdash \quad P \to R$$

where Q, P, R are propositions with truth values belonging to the set $\{True, False\}$. After simple manipulation the last rule may be rewritten:

$$\mathbf{P1}: \quad \neg P \vee Q$$

$$\mathbf{P2}: \quad \neg Q \vee R$$

....

$$\mathbf{P3}: \quad \neg P \vee R$$

where **P1** and **P2** are named premises and **P3** is the conclusion. In this form it is easy to generalize the resolution rule to fuzzy logic. In this case P, Q and R have truth values distributed in the interval $[0, 1]$ and we have:

$$\mathbf{P1}: \quad \max(1 - \chi_P, \chi_Q)$$

$$\mathbf{P2}: \quad \max(1 - \chi_Q, \chi_R)$$

....

$$\mathbf{P3}: \quad \max(1 - \chi_{\mathbf{P1}}, \chi_{\mathbf{P2}}).$$

The last form of the resolution principle allows us to carry out inference also in the case of non-deterministic or linguistic premises. For example this form of the resolution principle could allow us to express inferences like this:

If the spectrum follows an exponential law there is little-evidence *that the background is diffuse;*

*   if the background is diffuse there is* little-evidence *that the universe is steady;*

*   ...;*

*   if the spectrum follows an exponential law is* not evident *that the universe is steady.*

Knowledge based systems, that include fuzzy logic and its application to the design of inference engine, have been designed and implemented. The resolution principle in the context of fuzzy logic has been applied in the system RIMA (Di Gesù and Maccarone, 1988a), designed for the analysis of sparse images

The system CLASSIC (Thonnat, 1985) has been realized for the automatic classification of celestial objects, using also the fuzzy evaluation of the rule set.

11.4 Implementation Features

In this section, we give two examples of the use of fuzzy sets in order to represent non-deterministic knowledge and to handle mathematically imprecise variables in KBSs.

11.4.1 Non-Deterministic Decision Tables

One of the underlying principles of software engineering is that the methodology for solving a problem is based on techniques that are application-independent. In designing KBSs this rule must be followed closely. For example the internal representation of the knowledge must be easily updated to a new application or to a new situation. Also decisions must be data driven as well as the strategies to solve problems.

One of the most common techniques is based on the use of decision tables (DTs). These have been widely used in all phases of software engineering from system planning, through the software design process, down to software maintenance. Their use seems to be also effective in KBSs dedicated to data analysis problems. DTs may contain information about the knowledge base, predefined conditions on the status of this information and control of actions (procedures, decisions, conclusions).

DTs may be composed of four major parts: the condition *stub* and *entry*, and the action *stub* and *entry* (see Figure 11.5). The condition *stub* contains a row for each condition to be evaluated, the action *stub* contains a row for each action.

Condition Stub	Condition Entry
Action Stub	Action Entry

Fig. 11.5. Parts of a decision table.

The condition and the action *entry* sections are divided into columns called rules. Each column specifies values for conditions and resulting actions to be taken, whenever the conditions meet the specified values. In deterministic DTs the condition entries have truth values in the domain $\{T, F\}$.

In Figure 11.6 an example of a DT to sort N records is shown. The problem concerns the optimization of the sorting procedure on the basis of data organization and cardinality.

According to rule 5 of this table, if the number of records is greater than 10, one does not want to alphabetize, and recursion is not available; one then

	1	2	3	4	5	6
No.of records ≤10	T	T	F	F	F	F
Records have ≤3 fields	T					
Records Have >3 fields		T				
Records have >100 fields		F	T			
Alphabetizing				T	F	F
Recursion available					F	T
Call Insertion Sort	X					
Call Selection Sort		X				
Call Heapsort					X	
Call Quicksort						X
Call Bucket Sort				X		
Sort Array of link to rec.			X			
GO AGAIN			X			
EXIT	X	X		X	X	X

Fig. 11.6. Decision Table for sorting N records.

calls the Heapsort procedure to sort the records. If, however, all the above is true except that recursion is available, then rule 6 tells us to call the Quicksort procedure.

This example shows the flexibility and modularity of the DTs in representing decision trees in KBSs. However they can handle, only, deterministic models and may require a great amount of memory to represent very complex situations, where the number of conditions is too large to describe all possible conditions.

This weakness becomes dramatic in the case of KBSs dedicated to data analysis, where non-deterministic conditions are often used and subjective reasoning is dominant. As we have seen, fuzzy logic provides a mathematical basis to manage uncertainty in KBS. In this case the values of the *entry* conditions are non-deterministic, and may contain linguistic variables (low, small, high, large, very large, etc.) and/or modal operators (possible, necessary).

Non-deterministic decision tables (NDTs) were introduced in order to handle imprecise conditions (Francioni and Kandel, 1983), using the concept of fuzzy set theory. The conditions were defined as fuzzy variables, the rules then became combinations of these variables represented as fuzzy switching functions. In Figure 11.7, a NDT is shown for sorting optimization. In the table the first condition has been generalized, while the second, third and fourth conditions have been grouped. The table is now more representative of how people think, rather then being based on how the computer works.

It must be noted that NDTs require extra memory in order to store the tables required to represent the linguistic variables used in the *entry* condition.

In order to get a feel for how subjective evaluation in NDTs is, a Sensitivity Index (SI) is defined as follows:

$$SI = \frac{N_C N_R}{\sum M T_R}$$

No.of records	Low	Low Med	High Very	High	High	High
Size of records	Small	to large	large			
Alphabetizing				T	F	F
Recursion available					F	T
Call Insertion Sort	X					
Call Selection Sort		X				
Call Heapsort					X	
Call Quicksort						X
Call Bucket Sort				X		
Sort Array of link to rec.			X			
GO AGAIN			X			
EXIT	X	X		X	X	X

Fig. 11.7. Non-deterministic Decision Table for sorting N records.

where N_C is the number of non-deterministic conditions; N_R is the number of rules dealing with non-deterministic conditions; M is the number of possible values of a condition; and T_R is the total number of rules in the NDT.

From this definition it follows that SI = 0 implies that the NDT = DT, while SI = 1 implies that the NDT is completely non-deterministic. For the NDT in Figure 11.7:

$$SI = \frac{2 \times 6}{(1 + 1 + 2 + 2) \times 6} = 0.33$$

11.4.2 Fuzzy Arithmetic and Data Analysis

Fuzzy numbers may be considered in many ways. One of the most accepted definitions is to consider a fuzzy number as an extension of the concept of confidence interval (Kaufmann and Gupta, 1985). One of the motivations for such a choice is connected with the meaning of precision in measurements (errors). One of the merits of fuzzy set theory resides in the formalization of the algebra of fuzzy numbers, allowing the design of more effective algorithms for the handling of uncertain data.

Fuzzy arithmetics have been applied to the handling of experimental database management systems. For example in the implementation of a generalized equijoin operation, deterministic equality is generalized by the concept of approximate equality. Let A and B be real numbers; then:

$$A \simeq B \Longleftrightarrow [A - \epsilon(A), A + \epsilon(A)] \cap [B - \epsilon(B), B + \epsilon(B)] \neq \emptyset$$

where $\emptyset$ is the empty set, and $\epsilon(A)$ and $\epsilon(B)$ are the error bars in the evaluation of B. This definition may be rewritten in the formalism of fuzzy sets; let χ_A and χ_B be the fuzzy distribution related to "A" and "B", then:

$$A \simeq B \Longleftrightarrow \max\{\min\{\chi_A \text{ and } \chi_B\}\} > \alpha$$

with $0 < \alpha \le 1$.

The value of α may be settled on the basis of subjective knowledge and expertise. Usually the property of convexity is required in the definition of the membership function.

Fuzzy numbers are again useful for representing vague quantities such as *many, large, moderate, few*. Such terms are used, on occasion, by data analysts in preliminary analysis, and should be included in KBSs to help scientists to start the exploratory analysis.

Arithmetic operators have been introduced on fuzzy numbers (Dubois and Pradé, 1979b), such as addition, subtraction, multiplication and division, and their properties studied from the algebraic point of view.

Consider for example the representation of the fuzzy number *many*:

$$many : \textbf{integer} \rightarrow [0,1]$$

$$\chi_{many} =$$

$$\{1/.1, 2/.2, 3/.3, 4/.4, 5/.5, 6/.6, 7/.7, 8/.8, 9/.9, 10/1, 11/.9, 12/.8, 13/.7\}$$

and the representation of the fuzzy number *few*:

$$few : \textbf{integer} \rightarrow [0,1]$$

$$\chi_{few} = \{1/1, 2/.9, 3/.8, 4/.7, 5/.6, 6/.5, 7/.4, 8/.3, 9/.2, 10/.1\}.$$

The two numbers express the two propositions "approximately 10" and "approximately 1". Their sum, computed according to the definition of fuzzy addition, is another fuzzy number:

$$(many + few)(n) = \max_{n=u+v} \{\min\{many(u), few(v)\}\}.$$

The last number is equivalent to the proposition "approximately 11"; in fact:

$$(many + few)(11) = \max\{1, .9, .8, .7, .6, .5, .4, .3, .2, .1\} = 1$$

$$(many + few)(10) = \max\{.9, .8, .7, .6, .5, .1, .2, .3, .4, .5\} = .9$$

$$(many + few)(1) = \max\{0, 0\} = 0$$

$$(many + few)(2) = \max\{.1, .1\} = .1.$$

This example shows that the fuzzy number "*many + few*" is a little more than the fuzzy number "*many*".

11.5 Fuzzy Sets and Image Analysis

Fuzzy indicators may be applied in texture classification problems (Di Gesù and Maccarone, 1988b). These consist of partitioning an image, X, by means of an equivalence relation defined on its pixels. The equivalence classes are also named segments. In the case of *Sparse Images* a single link clustering algorithm has been applied to perform the segmentation. Two pixels belong to the same region *iff* their distance is less than a given positive threshold, ϕ. The best threshold is then chosen on the basis of the fuzzy indicator:

$$\mathbf{I}_\phi = 1 - \frac{1}{N_r} \sum_i \mathbf{Q}(m_i; \phi)$$

Here $\mathbf{Q}(m_i; \phi)$ is the probability that the region $R_i(\phi)$ has m_i pixels, and N_r is the number of regions. It may be interpreted as the degree of membership of the segment $\{R_j(j)\}$ in the fuzzy set **j**-segmentation of the Sparse Image. The *best threshold* is then determined as follows: *Choose ϕ' iff* $\mathbf{I}_{\phi'} = max\{\mathbf{I}_\phi\}$.

This method is very effective whenever the background is structured, or the pixel intensities have a wide smooth variation in segments and image intensities are very low (of the order of 1 count/pixel).

Image classification may also be accomplished using fuzzy techniques. For this purpose, the normalized quadratic axial moments (NQAM) are computed in r directions. The parameters NQAM are related to the object symmetry in a similar way to Fourier descriptors. Each object, x, is represented in an r-dimensional space:

$$x(m_1, m_2, \ldots, m_r) \qquad \text{and} \ \ 0 \le m_i \le 1$$

where m_i is the NQAM computed in the q_i direction. NQAMs are then used to build a membership function of an object, x, with respect to a fuzzy class, k. This membership function is $\chi_k(x)$. An element, $I^{(i)}$, such that $\chi_k(I^{(k)}) = 1$, is said to be ideal and may be determined by assuming a model of the shape or by a training set algorithm. The degree of membership of x in a fuzzy class k may be defined as:

$$\chi_k(x) = 1 - \frac{1}{r} \sum_{i=1}^{r} | m_i - I_i^{(k)} |$$

or by using the normalized entropy function (De Luca and Termini, 1972):

$$\chi_k(x) = 1 - \frac{1}{r} \sum_{j=1}^{r} [-\eta_{kj} \log \eta_{kj} - (1 - \eta_{kn}) \log(1 - \eta_{kj})]$$

where

$$\eta_{kj} = | m_j - m_j^{(k)} | \le 1.$$

Both of these may be interpreted as the degree of belonging to class k. The classification rule may be based on the maximization of $\chi_k(x)$ over the number of classes.

This method has been tested on a sample of preclassified galaxies belonging to the classes: { ellipse, circle, S-2 arms, S-4 arms }, for a total of 600 events.

The density, ρ, of the "on" pixels was very low (0.03 counts/pixel $\leq \rho \leq$ 0.09 counts/pixel). The classification rate versus the pixel density is shown in Figure 11.8.

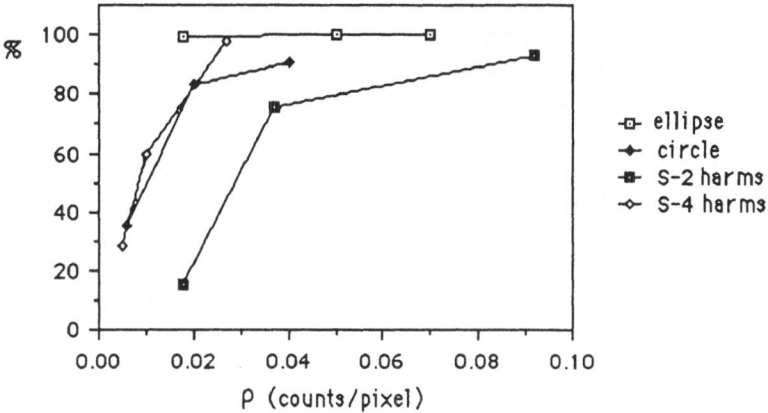

Fig. 11.8. Classification rate versus pixel density.

11.6 Concluding Remarks

Fuzzy set theory may be applied in designing KBSs in which the need for the representation of uncertain or non-deterministic knowledge exists. The theory is quite young and not completely well understood. It has been introduced to overcome some of the conceptually open problems of probability theory (how to define or estimate the pdf of small data sets) and to quantify some aspects of human reasoning. The results of its application are often satisfactory; however only heuristic arguments support them. Still more work should be done in order to have formal proof of the power and consistency of this frontier theory.

References

1. Davis, R. (1980), "Application of meta level knowledge to the construction, maintenance, and use of large knowledge bases", *Knowledge Based Systems in Artificial Intelligence*, R. Davis and D.B. Lenat (eds.), McGraw-Hill, New York.
2. De Luca, A. and Termini, S. (1972), "A definition of a nonprobabilistic entropy in the setting of fuzzy set theory", *Information and Control*, **20**, 301–312.

3. Di Gesù, V. (1989), "Mathematical morphology and image analysis: a fuzzy approach", Workshop on *Knowledge Based Systems and Models of Logical Reasoning*, Cairo, Egypt.

4. Di Gesù, V. (1986), "Problems and possible solutions in the analysis of sparse images", *Pattern Recognition Theory and Applications*, P.A. Devijver and J. Kittler (eds.), Springer-Verlag, Berlin, 277–286.

5. Di Gesù, V. (1988), "Fuzzy sets and data analysis", *Astronomy from Large Database: Scientific Objectives and Methodological Approaches*, F. Murtagh and A. Heck (eds.), European Southern Observatory, Garching bei München, 183–197.

6. Di Gesù, V. and Maccarone, M.C. (1988a), "An aproach to random image analysis", *Image Analysis and Processing II*, V. Cantoni, V. Di Gesù and S. Levialdi (eds.), Plenum Press, New York, 111–118.

7. Di Gesù, V. and Maccarone, M.C. (1988b), "Vision problems in sparse images", *Machine Vision*, G. Pieroni (ed.), Springer-Verlag, Heidelberg.

8. Dimitrov, V. (1983), *Fuzzy Sets and Systems*, **9**, 25–40.

9. Dubois, D. and Pradé, H. (1978a), *International Journal of Systems Science*, **9**, 357–360.

10. Dubois, D. and Pradé, H. (1978b), *International Journal of Systems Science*, **9**, 613–626.

11. Dutta Majumder, D. (1980), *International Journal of Systems Science*, **11**, 1435–1445.

12. Dutta Majumder, D. (1986), *Fuzzy Mathematical Approach to Pattern Recognition*, Halsted Press, New York.

13. Einstein, A. (1916), *Annals of Physics*, **49**, No. 215.

14. Francioni, J.M. and Kandel, A. (1983), *Fuzzy Sets and Systems*, **9**, 41–68.

15. Galileo Galilei (1632), *Dialogo sopra i due massimi sistemi del mondo.* (*Dialogue Concerning the Two Chief World Systems*, translated by Stillman Drake, University of California Press, 1953, revised 1957.)

16. Kandel, P.F.A. (1976), *IEEE Transactions on Systems Man and Cybernetics*, **SMC-6**, 215–219.

17. Kandel, P.F.A. (1986), *Fuzzy Mathematical Techniques with Applications*, Addison Wesley, Reading, MA.

18. Katsaras, A.K. (1981), *Fuzzy Sets and Systems*, **6**, 100–111.

19. Kaufmann, A. (1980), *Theory of Fuzzy Sets and their Applications*, Academic Press, New York.

20. Kaufmann, A. and Gupta, M. (1985), *Introduction to Fuzzy Arithmetic*, Van Nostrand, New York.

21. McCalla, G. and Cercone, N. (1983), *IEEE Computer*, **16**, 12–18.

22. Michie, D. (1983), *Intelligent Systems: The Unprecedented Opportunity*, Ellis Horwood, Chichester, 57–69.

23. Negoita, C.V. (1985), *Expert Systems and Fuzzy Systems*, Benjamin/Cummings, Menlo Park, CA.

24. LaChat, M.R. (1986), *A.I. Magazine*, **7**, 70–79.

25. Lenat, D.B. (1983), "Eurisko: a program that learns new heuristics and domain concepts. The nature of heuristics III: Program design and results", *Artifical Intelligence*, **19**, 189–249.

26. Rescher, N. (1968), *Topics in Philosophical Logic*, D. Reidel, Dordrecht.
27. Rosenfeld, A., Hummel, R.A. and Zucker, S.W. (1976), *IEEE Transactions on Systems, Man, and Cybernetics*, **SMC-6**, 420–423.
28. Rosenthal, D.A. (1988), "Applying artificial intelligence to astronomical databases: a survey of applicable technology", *Astronomy from Large Databases: Scientific Objectives and Methodological Approaches*, F. Murtagh and A. Heck (eds.), European Southern Observatory, Garching bei München, 245–259.
29. Serra, J. (1982), *Mathematical Morphology*, Academic Press, New York.
30. Shortliffe, E.H. (1976), *Computer Based Medical Consultations: MYCIN*, Elsevier, New York.
31. Stefik, M. (1986), *A.I. Magazine*, **7**, 34–46.
32. Thonnat, M. (1985), "Automatic morphological description of galaxies and classification by system", Internal Report, INRIA, No. 387.
33. Thonnat, M. and Clément, V. (1989), "OCAPI: an artificial tool for the automatic selection and control of image processing procedures", *Data Analysis in Astronomy III*, V. Di Gesù, L. Scarsi, P. Crane, J.H. Friedman, S. Levialdi and M.C. Maccarone (eds.), in press.
34. Yager, R.R. (1980a), *Journal of Cybernetics*, **10**.
35. Yager, R.R. (1980b), *Journal of Cybernetics*, **10**.
36. Ullman, J.D. (1982), *Principles of Database Systems*, 2nd ed., Computer Science Press, Rockville, MD.
37. Waterman, D.A. (1986), *A Guide to Expert Systems*, Addison-Wesley, Reading, MA.
38. Woods, W.A. (1975), "What's in a link: foundation for semantic networks", in *Representation and Understanding: Studies in Cognitive Science*, Bobrow and Collins (eds.), McGraw-Hill, New York, 35–82.
39. Zadeh, L. (1962), *Proceedings of the Institute of Radio Engineers*, **50**, 856–865.
40. Zadeh, L. (1985), *IEEE Transactions on Systems, Man, and Cybernetics*, **SMC-15**, 754–763.

12 An Approach to Heuristic Exploitation of Astronomers' Knowledge in Automatic Interpretation of Optical Pictures

A. Accomazzi (1), G. Bordogna (2), P. Mussio (1) and A. Rampini (2)

(1) Università degli Studi di Milano
Dipartimento di Fisica
Via Viotti 5
I-20133 Milano
Italy

(2) Istituto di Fisica Cosmica e Tecnologie Relative, C.N.R.
Via Ampere 56
I-20131 Milano
Italy

12.1 Introduction

The visual interpretation of astronomical images is based on heuristic (Pólya, 1945) procedures, which complement precise quantitative measurements. Examples from the literature of such heuristics are the measurement of the diameter of star images in sky survey plates (King and Raff, 1977), galaxy identification (Godwin et al., 1983), and the classification of galaxies based on the recognition of object features such as bars or arms (Thonnat, 1985; Balestreri et al., 1979) according to the Hubble "tuning fork" even in its more extensive version (Hubble, 1936; Sandage and Sandage, 1983).

In these cases, criteria such as "we estimated the diameter of each star image at a point where the image was a fairly dark gray but not quite dark" (King and

Raff, 1977) are necessarily assumed in the visual classification. This kind of plausible reasoning (in the sense of Pradé, 1985) depends upon a visual judgement and is based on astronomers' expertise and on uncertain rule reasoning: observational data become constraints that the deduced results, which are therefore subjective, must satisfy. Different astronomers may yield different classifications of the same object, and since this is a well known situation, cross examinations of the data and discussions of the cases of disagreement are used to reach consistent results (see for example Godwin et al., 1983, and King and Raff, 1977). When dealing with the design of automated systems for astronomical image interpretation, the problem of translating these heuristics into executable programs arises. In this paper we argue that such problems can be faced by the combined use of certain generative rule-based devices, the Conditional Attributed Lyndenmaier Systems (CAIL; Mussio et al., 1989), which allow the description of structures present in an image, and of Multi-Valued Logic Trees (MVLT; Garribba et al., 1985), which yield the judgement of an examined structure from its previously obtained description. These tools allow the definition of algorithms, which mimic the astronomer's activity and approximate the heuristic procedure used. In order to obtain this definition, the astronomer's knowledge about the problem must be studied and a precise description of the image characteristics, called the astronomer's image model (Ahuja and Shachter, 1981), that he/she seems to exploit in the design and execution of image interpretation, must be defined. For this reason, before introducing the proposed interpretation strategy, the image model is discussed, and its influences on the definition and implementation of the tool is outlined. On the whole, our proposal is an attempt to formalize and translate into programs the visual classification of tracks in astronomical images. This approach, if successful, complements methods which have been proposed for classification objectives with quantitative data (Murtagh and Lauberts, 1986; Thonnat, 1985). Furthermore, some of these can eventually be proposed as semantic rules and even used as decision criteria by our system, since the aim of the attributed generative approaches is the blending of quantitative and structural methods (Fu, 1982).

12.2 The Image Model

The image model provides a precise description of the image characteristics necessary for an efficient design of the operations (Ahuja and Shachter, 1981) which lead to image interpretation. It is an approximation of the knowledge that the astronomer exploits in his/her activity of visual interpretation.

In the proposed approach, the image model is studied at three different levels, which reflect the different levels of agreement among astronomers about the process being modelled. They are: the experiment model, i.e. a model of how astronomers deal with those situations in which theories about celestial objects are to be exploited in the interpretation of data; the scene model, describing the set of object types and relations the astronomer can expect to observe in the scene; and the imaging process model, i.e. the model of the process involved in image formation and acquisition.

Some considerations about the scene and imaging process model (ip model in the following) are necessary.

Since an image is a 2D projection from a unique direction of sight of a 3D scene, some objects may be occulted to the observer by others and structured objects can be characterized with difficulty when they are not face-on. Images obtained from the observation of this kind of scene may miss the track of some objects because of occultation, limited resolution or low surface brightness. Such images can be defined as being multiresolved, because faint tracks of objects as well as resolved ones may appear.

In other words, the interpreter often deals with partial data about the object under classification because of the inherent nature of the scene.

As to the ip model, astronomical images are characterized by intrinsic noisiness, which depends on the image acquisition process. Certain characteristics of the ip model are common to large classes of observations. For example, stars and galaxies observed from ground-based devices result in tracks with fuzzy boundaries. Other characteristics depend on the observational device. For example, plates are affected by film-grain noise, which must be estimated if the detection of faint objects is one of the goals of the interpretation. Due to the fact that film grain noise can be known only approximately (Andrews and Hunt, 1977), the visual detection of faint objects will not only involve imaging considerations, but will be based also on the scene knowledge and on the astronomer's point of view about the nature of the object.

On the other hand, all three models are exploited when the astronomer denotes with the name of a celestial object a track in the analogical images. Some of these tracks are labelled with definitive labels, such as "galaxy", some with labels which express uncertainty in judgement, such as "possible galaxy" (Godwin et al., 1983). In other words, when looking for a galaxy an astronomer has to judge if a track in the image has some properties such that it can be considered an element of the set named "galaxy". When it is possible and necessary, the astronomer states even the (informally defined) plausibility of his/her verdict. When dealing with digital or digitized images, the image model is more constrained, because a digital image is a finite set of coloured pixels. Let the name "structure" denote a set of pixels in the image, which share some common geometric, topological, or colour properties. By each label, say "galaxy", the astronomer denotes now a large but finite set of different structures which may appear in the finite digital image. This set can be recognized on the basis of its properties and context as in the analogical procedure associating with it, when it is possible, a plausible verdict about its nature.

12.3 An Overview of the Proposed Interpretation Strategy

From the computer science point of view, every finite set of structures, that the astronomer implicitly associates with a name such as "galaxy", is a bidimensional language (Fu, 1982; which will be formally defined later). As a language, it can be described by a generative device (Salomaa, 1973) which operates on a bidimensional vocabulary, i.e. a set of structures. The problem of finding a structure of interest for the astronomer is reduced to the solution of the membership problem stated as: given a finite structure, decide, in a finite number of steps, whether or not it is a member of a set, described by the generative device.

In the case of astronomical images, the algorithm proposed to mimic the astronomer in the solution of the membership problem is divided into three steps. In the first two stages the total or partial description of the digital image structures is performed, in the last the structures are given the "meanings" with reference to the real scene.

In the first description stage the Coloured Digital Image (CDI) is segmented, that is the structures which are candidates to be meaningful are selected by evaluating some propositions

$$\mathbf{P} : P(x, y, c, X_1, Y_1, C_1, \ldots, X_n, Y_n, Z_n, H, H_1, \ldots, H_k)$$

where:

x, y are the positional coordinates of a generic pixel p,

c is its colour,

X_i, Y_i and C_i are the positional coordinates and colour of its i'th neighbour,

H_j is generic data external to the image.

In our case $\mathbf{P}$ may formalize both precise statements such as "the colour of meaningful pixels is above a given, locally variable threshold thr", or equivalent linguistic expressions such as "bright source", translated into: $\mathbf{P} : c > thr(x, y)$ where c is the colour of pixel p whose positional coordinates are x, y; or even "faint source which appears homogeneous within the texture of its context" which can be translated into a more complex algorithm which will be discussed in section 12.3.3.

The result of the application of $\mathbf{P}$ to all pixels in the CDI (Fig. 12.1a), is a Binary Digital Image (BDI; Fig. 12.1b). In a BDI, pixels whose characteristics satisfy $\mathbf{P}$ are set to 1 and are called "visible" or "candidate" (to belong to a meaningful structure). The whole set of visible pixels is called the foreground of the image, the set of invisible ones the background.

In the second stage, the foreground is described, looking for the presence in it of those characteristic structures which in the astronomer's definition are candidates to be tracks of celestial objects (BDI description). Structures are identified and to each one a description, i.e. a set of observed topological, geometrical and shape properties, is associated following the astronomer's definitions, codified into CAILs which will be defined in the next section. Each property is evaluated as it is a possible hint of the nature of the track (real source or noise) from which the structure derives.

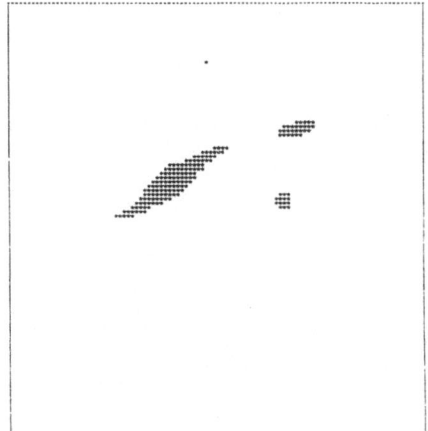

Fig. 12.1. (a) Subfield from cluster of galaxies of the southern sky near $\alpha = 4h\ 30m$, $\delta = -54°$. Plate 157J of the ESO Sky Survey; scale $= 67"$/mm, magnitude limit $= 20$–21; digitized 20'20"x20'20" (1000x1000 pixels). Digitization: ESO Garching microdensitometer. (b) BDI obtained by thresholding the image in (a).

The set of candidate pixels corresponding to a single object need not be, and in general is not, a simple connected set, because for example, tracks of different objects may be overlapping or a structure may not display all the features characterizing an object type. Therefore this stage is aimed at searching for hints and clues which indicate the possible existence of a given object track. To this end the binary structures in the image are described in terms of their components (convexities, concavities, inlets, outlets) which may be the hints of the presence of a given feature, say an arm, a clue for the subsequent interpretation. So, for example, concavities and convexities may be considered a hint which indicates the possible beginning of an outlet, a structure springing out from a blob (Fig. 12.2). If a concavity (convexity) is found, the search for an outlet starts.

If such a structure is recognized and its area is above a given threshold, its sides are studied, and their characteristics are deployed. These hints and clues are thereafter judged, taking into account the context in which they appear. So if the sides are nearly straight convergent segments, the structure is a candidate to be a "pattern diffraction spike" or an arm of galaxy. On the contrary if they are curved and slowly convergent, it could be the track of an arm of a spiral galaxy (Fig. 12.3).

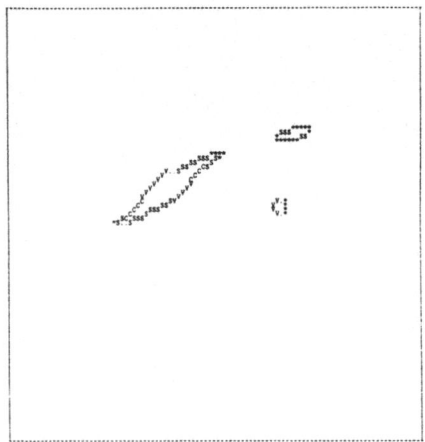

Fig. 12.2. The structures in Fig. 12.1 (b) are described in terms of elementary features present on their contour. 'c' denotes a concavity, 'v' a convexity, 's' a stepwise side, '*' a convex end, '.' all the remaining contour elements.

The judgment is established by a two step procedure. First each attribute of the identified structures is judged as if it were the only observation about the unknown object by a labelling function (L-function), then the use of MVLTs allows the exploitation of contextual relations among the attributes in order to plausibly classify the structures as tracks of objects of specified types possibly present in the scene.

The three stages outlined are not watertight. Results reached in one step may compel the repetition of a preceding analysis. In this case, however, the second analysis obviously exploits knowledge about the cause of the failure of the preceding analysis. This technique has been applied in fields other than astronomy (Della Ventura et al., 1987) and to the interpretation of astronomical digitized plates by one of the authors (Mussio, 1985). A more advanced architecture, born of a collaborative effort, is discussed in Kurtz (1989).

In this paper, we focus on how the knowledge of the astronomer (i.e. the image models) is codified into formal tools, suitable to be programmed.

To this end, first we introduce the simpler problem of BDI description, then we extend the method to 2D description tools which are used in image segmen-

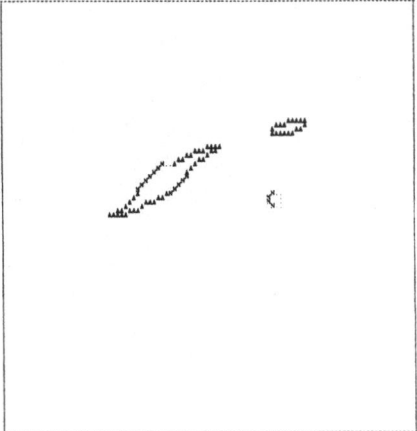

Fig. 12.3. High level description of the contours shown in Fig. 12.2. 'A' denotes the contour of a POSSIBLE ARM, 'N' of a POSSIBLE NUCLEUS or BULGE, '.' all the remaining contour elements.

tation and conclude with a quick look at the judgement phase by L-functions and MVLTs.

12.3.1 Binary Image Description

In the stage of Binary Image description, the shape of structures is described, looking for features (set of pixels belonging to the structure), which are of interest for the astronomer.

The structure description in a BDI starts by coding each visible pixel into a decimal number:

$$\text{CODE} = \sum C(i)\, 2^i$$

where i is the order number of the 8-neighbours, zero being associated with the left hand side neighbour on the same row, the others being enumerated clockwise (Merelli et al., 1985), and $C(i) = 0$ if the i'th neighbour is visible.

In order to make the descriptions based on the code more readable, the code numbers are often translated into mnemonic symbols. A pixel of the image is so described by a symbol and its positional coordinates. Such a triple is called an *attributed symbol*. In Merelli et al. (1985) it is demonstrated that the foreground can be described by the enumeration of its *Multiple Elements* (MEs), a subset of the contour pixels. An ME is a visible pixel whose neighbourhood contains a corner, i.e. a sequence of invisible pixels with at least one element in an even coded position, such that the number of invisible elements in the sequence is different from 3 (the corner is not a straight one). In Table 12.1 the set ALF of

Table 12.1. The set of MEs.

$$ALF = \{\ A,B,C,D,E,F,G,H,I,J,K,L,M,N,O,P,Q,R,S,T,U,V,$$
$$W,X,Y,Z,\underline{A},\underline{B},\underline{C},\underline{D},\underline{E},\underline{F},\underline{G},\underline{H},\underline{I},\underline{J},\underline{K},\underline{L},\underline{M},\underline{N},\underline{O},\underline{P},\underline{Q},\underline{R}\ \}$$

CODE	SYMB	CODE	SYMB	CODE	SYMB	CODE	SYMB
1	A	4	B	16	C	64	D
3	E	6	F	12	G	24	H
48	I	96	J	192	K	129	L
15	M	30	N	60	O	120	P
240	Q	225	R	195	S	135	T
31	U	62	V	124	W	248	X
241	Y	227	Z	199	$\underline{A}$	143	$\underline{B}$
63	$\underline{C}$	126	$\underline{D}$	252	$\underline{E}$	249	$\underline{F}$
243	$\underline{G}$	231	$\underline{H}$	207	$\underline{I}$	159	$\underline{J}$
127	$\underline{K}$	254	$\underline{L}$	253	$\underline{M}$	251	$\underline{N}$
247	$\underline{O}$	239	$\underline{P}$	223	$\underline{Q}$	191	$\underline{R}$

codes and corresponding symbols of the MEs is shown. This set is an alphabet and is denoted by the word ALF from "ALFabeto" in Italian.

A description represented as a string of attributed symbols is called a *linear representation* (lr). A lr is *congruent* if each couple of consecutive symbols denotes two D-associated MEs, i.e. if their codes and positions are such that they are the extreme elements of a segment, a sequence of contour pixels in which no change of direction occurs. A useful congruent description is called the *primal* one, in which each contour of a connected set of pixels is described by its congruent lr. In Merelli et al. (1985) an algorithm which obtains the primal description from any BDI is described. Given the primal description, the descriptions of the structures of interest for an experiment can be derived by defining the set of pixels composing them, the associated set of attributed symbols and the mapping which translates one description into another. To this end, first the structures which are of interest in the astronomical analysis (features) are characterised as a set of MEs, then a symbol is used to denote their type. For example a concavity is defined as a set of at least two MEs which are D-associates and are vertices of a corner of less than three pixels. These elements are named *Local concavities* and the symbol denoting them is Lconc. A concavity is therefore composed of a set of D-associated Local concavities; the symbol denoting this type of structure is CONC. The concavity attributes which are of interest for galaxy and star recognition (the curvature SLOPE, the starting START and ending END coordinates of the concavity) are identified by the astronomer, and their range is established. SLOPE is a variable, whose value ranges in $\mathbb{Z}$. Such a value can be computed from a property of the component MEs, called the Local-Slope. Both START and END are variables described by a couple of coordinates (x,y) with $1 \le x \le M,$ $1 \le y \le N$ where $M, N \in \mathbb{N}$ are the numbers of rows and columns in the CDI, respectively. Therefore a concavity in a BDI is described by the symbol CONC and by the numbers (named the concavity properties) which are the values of its SLOPE, END, START attributes for that particular structure. The quadruple constituting this description is again an attributed symbol.

The set of all the attributed symbols, which are necessary for building the image descriptions, is called an attributed alphabet. In the example, the alphabet for the concavity detection is constituted by

$$V = \text{ALF} \cup \{\text{Lconc}, \text{CONC}\}$$

Generally, one can say that, to describe a structure, first a set V of symbols is established, then for each symbol $v \in V$ a finite set of attributes $A(v)$ is defined; each attribute $a_V \in A(v)$ has a finite or infinite set of possible values $D_V(a_V)$, called the set of possible properties of v.

Once a feature and the attributed symbol associated with it is identified, a function must be defined to allow the recognition and the description of the feature from the descriptions of its components. This function is described by a generative device from the L system family (Salomaa, 1973). An L system maps an lr into a new lr in which a unique attributed symbol denoting the sought feature synthesizes the attributed symbols of the components. L-systems are parallel generative devices used to define formal languages in an evolutionary way; i.e. they allow the defining of how a string in a language evolves into another string of the same language, rather than describing the deep structure of a string as grammars do (Salomaa, 1973). The L systems which allow interaction among symbols, i.e. the use of contextual rules, are called IL systems. In Mussio et al. (1989) Conditional Attributed IL systems (CAIL) were introduced; they extend IL systems to treat strings of attributed symbols in a conditional way. An example of a CAIL, able to recognize concavities in a primal description, built according to the rules exposed in the next section, is shown in Table 12.2.

12.3.2 CAIL and BL Systems

A conditional attributed IL-system (CAIL) is a quadruple

$$\textbf{CAIL}: \; < \textbf{V}, \textbf{F}, \textbf{Ax}, \textbf{Mr} >$$

where: V is an attributed alphabet (a finite set of attributed symbols); Ax is a set of words over V, named the "set of axioms"; F is a set of contextual conditional attributed productions; and Mr is a set of metarules which specifies how to use the rules in F to derive an axiom from a given string P on V.

Each production in F is divided into two parts: a contextual conditional syntactic rule and a semantic rule. Contextual conditional rules allow the handling of a situation in which the rewriting of a sequence of symbols can depend both on the properties of their attributes and on the context in which the sequence appears. A contextual conditional syntactic rule has the form

$$c_1 \circ X \circ c_2 \dashv C \; s \mapsto - \circ Y \circ -$$

Table 12.2. Concavity definition rules.

$V = ALF \cup \{$ Lconc, CONC $\}$

$Ax \in V^*$

F set of rewriting rules: $F = \{$ F1, F2, F3 $\}$

a) Syntactic part:

    F1: $x1 \circ x2 \rightarrow$ Lconc $\circ$ -

    F2: $x1 \circ x2 \circ \beta \rightarrow \cdot \circ$ Lconc $\circ$ -

    F3: $\alpha \circ ($Lconc$)^n \circ \omega \rightarrow \cdot \circ$ CONC $\circ$ -

    $x1 \circ x2 :=$ EF/FG/GH/HI/IJ/JK/KL/LE/

                FE/GF/HG/IH/JI/KJ/LK/EL/

                FA/HB/JC/LD/KA/EB/GC/ID/

                AF/BH/CJ/DL/AK/BE/CG/DI/

                AB/BC/CD/DA/BA/CB/DC/AD

    where α , β , $\omega \in$ ALF - { A,B,C,D,E,F,G,H,I,J,K,L } , $n > 1$

    := means 'is equivalent to' and / means 'or'

b) Semantic part:

    $A($Lconc$) = \{$ Local-slope, Coord $\}$

    $D($Local-slope$) = \{$ -2,-1,1,2,3,4 $\}$

    $D($Coord$) = MxN$

    $A($CONC$) = \{$ SLOPE, START, END $\}$

    $D($SLOPE$) \subset Z$

    $D($START$) = D($END$) = MxN$

    F1: Local-slope(Lconc) = Local-slope($x1$)

       Coord(Lconc) = Coord($x1$)

    F2: Local-slope(Lconc) = Local-slope($x2$)

       Coord(Lconc) = Coord($x2$)

    F3: SLOPE(CONC) = $\sum_i$ Local-slope(Lconc$_i$)

       START(CONC) = Coord(Lconc$_1$)

       END(CONC) = Coord(Lconc$_n$)

    where Coord(x) refers to the positional coordinates of the ME named x

          Local-slope(x) is the local slope of the ME named x

          M and N are the number of rows and columns of the digital image, respectively

Each rule is identified by a label F followed by a number, which allows the coupling of its syntactic and semantic parts.

where $X = x_1 \circ x_2 \circ \ldots \circ x_n$, $Y = y_1 \circ y_2 \circ \ldots \circ y_m$, $s = s_1 \circ s_2 \circ \ldots \circ s_k$, c_1 and c_2 are strings over V (c_1, c_2 can also be the empty string, λ) and $\circ$ is the symbol of concatenation.

The left part of the rule is called antecedent, the right part consequent. A hyphen "–" in a position of the consequent indicates that any string of symbols may occupy that position. The symbol $\dashv \mapsto$ stands for rewrite under a condition, where the condition is denoted by the expression C between the bars. C is the name of a function $C : s \rightarrow \{0,1\}$. The syntactic rule is applied only if $Cs = 1$.

Semantic rules allow to compute the values of the attributes associated with symbols involved in the rewriting. The semantic rule is a set of expressions of the form:

$$f : D\psi(\alpha_{1\psi}) \ldots D\psi(\alpha_{n\psi}) \ldots D\omega(\delta_{1\omega}) \ldots D\omega(\delta_{n\omega}) \rightarrow D\xi(\gamma_{1\xi})$$

where

$$\psi, \omega, \xi \in \{x_1, \ldots, x_n, y_1, \ldots, y_m\};$$

$$\alpha_{i\psi} \in A(\psi), \delta_{i\omega} \in A(\omega), \gamma_{i\xi} \in A(\xi).$$

The only metarule used here specifies how a string can be directly derived by another one.

Let $X = x_1 \circ x_2 \circ \ldots \circ x_n$ be a string over V. For each x_i in X, let

$$\text{Pref}(x_i) = x_1 \circ \ldots \circ x_{i-1}$$

$$\text{Suf}(x_i) = x_{i+1} \circ \ldots \circ x_n.$$

Let $W = w_1 \circ \ldots \circ w_t$ and $Z = z_1 \circ \ldots \circ z_t$ be words over V such that

$$W = u_1 \circ P_1 \circ \ldots \circ u_n \circ P_n \circ u_{n+1},$$

$$Z = u_1 \circ Q_1 \circ \ldots \circ u_n \circ Q_n \circ u_{n+1}$$

where $P_i = p_{i1} \circ \ldots \circ p_{in}$ and $Q_i = q_{i1} \circ \ldots \circ q_{im}$ $u_i \in V^*$ for $i = 1, \ldots, n+1$ and $p_{ij}, q_{ij} \in V$ for each i and j. W directly generates Z (in symbols $W \Rightarrow Z$) iff $\forall$ P_i the contextual conditional syntactic rule

$$c_{i1} \circ P_i \circ c_{i2} \dashv Cs \mapsto - \circ Q_i \circ -$$

is in F, $C_i s_i = 1$, $\text{Pref}(p_{i1}) = w_1 \circ \ldots c_{i1}$ and $\text{Suf}(p_{in}) = c_{i2} \circ \ldots \circ w_t$.

An attributed conditional IL-system LS generates the language

$$L(LS) = \{y \mid \exists \, x \in A_x, \exists \, n \in N, y \in L_n(x)\}$$

where $L_n(x)$ is defined thus:

$$L_0(x) = \{x\}, \quad L_{n+1}(x) = \{y \mid \exists \, z, \, z \in L_n(x), \, y \Rightarrow z\}.$$

The generation is from y to z due to the analytical (i.e. recognition) use of the CAIL-system (Salomaa, 1973).

The L system approach can be extended also to the bidimensional case defining bidimensional words and rules. In this case the L system is called a bidimensional L system (BLS). A bidimensional word on V is a matrix of dimensions $M \times N$ (with $M, N \in \mathbb{N}^+$) whose entries are in V:

$$\omega = \| w_{ij} \|, \quad w_{ij} \in V, \quad i = 1 \ldots M, \ j = 1, \ldots N.$$

To describe the evolution of a bidimensional word, a set of bidimensional rules is defined.

The set R_b of bidimensional rules on V is a collection of couples $< \alpha, \beta >$ (denoted as $\alpha \rightarrow \beta$) where α, β are bidimensional words on V of the same dimension.

The parallel bidimensional direct generation process is stated as follows: a bidimensional word η over V generates directly a bidimensional word ω, in symbols $\eta \Rightarrow_{R_b} \omega$, if each subword η_k of η is rewritten into a subword ω_k of ω, and $< \eta_r, \omega_r > \in R_b$. From this formalization, we can derive both the definition of parallel bidimensional generation $= * \Rightarrow_{R_b}$, which is as usual the reflexive transitive closure of $\Rightarrow_{R_b}$, and the definition of the bidimensional language L_b, which can be stated as before.

12.3.3 Coloured Image Description by BLS

The definition of BLS allows us to extend the description method outlined for the binary image to the segmentation stage (Accomazzi et al, 1989). Here again, the use of attributed generative systems allows us to exploit heuristic criteria, based on different astronomers' models. That is, we use these tools mainly when properties based on linguistic definitions have to be algorithmically expressed. This is often the case of those pixels which are candidate to belong to structures of interest, that are all the excesses from the image background plausibly associated with a signal. The identification of these structures is performed by looking at the CDI as a 3D digital surface, in which the candidate tracks appear as hills emerging from a region characterized by randomly oriented short valleys and ridges. An emerging structure is always characterized by the presence of a maximum plane, a set of one or more pixels of the same colour, surrounded by a slope region, a set of pixels with a common growth direction of colour towards the maximum region. The emerging structure identification is achieved through the definition of a set of "positive" structures, the maximum planes and slope regions, and of a set of "negative" ones, the unstructured noisy regions constituted by random short valleys and ridges.

For pixel $p : (x, y, c)$ the "Slope Code" SC is defined as

$$SC = \sum | C(i) > C | \ 2^i$$

where i is the order number of the 8-neighbourhood, as previously defined and $C(i)$ is the colour of the i'th neighbour of pixel p. Each possible combination of growth directions of colour is described by a suitable SC.

The set of bidimensional words of interest is thereafter defined through a set of BLSs used in cascade (i.e. the axioms of one BLS are the recognized words of the preceding one). In this phase our approach is conservative: the set of candidate structures which possibly correspond to the excesses in the CDI, is defined so as to try to minimize the loss of meaningful pixels, although some noisy structures are selected.

Some structures which were found to be of interest are:

- Maxima: sets of pixels with SC = '0', the non-maximum neighbours of which show a growth towards the '0' coded pixels.
- Minimum and flex planes: sets of pixels with SC = '0' which are not maxima.
- N_Slopes: sets of pixels whose SC denotes monotonic growths in N adjacent directions ($N = 1, \ldots, 4$).
- Ridges and Valleys (RiVas): sets of pixels whose growth directions cluster into two connected sets. They may be single ones if constituted by one pixel, multiple ones if constituted by more pixels. These latter pixels may individually belong to other classes.
- Gullies: single pixels whose SC denotes 5 adjacent growth directions.
- Bottom Valleys: single pixels whose SC denotes at least 6 adjacent growth directions.

All BLSs used share a common alphabet, $V = V_1 \cup V_2$, where V_1 is the set of primitive slope codes as previously defined, V_2 is a set of additional codes used to label the different structures recognized step by step,

$$V_1 = \{0, \ldots, 255\},$$

$$V_2 = \{256, 257, 258, 259, 260, 261, 300, 301, 302, 303, 400, 401, 402\}$$

and a common interpreter Mr. An example of one of the BLSs is shown in Fig. 12.4.

This BLS identifies valleys of any length lying on straight lines from right (EAST) to left (WEST) of an image. Similar sets of rules allow the identification of valleys in any direction. Meandering valleys can be obtained by the combination of linear ones.

Valleys whose length is above a given threshold are used to detect the boundary regions separating overlapping excesses from the background. Short valleys are features of the background, and their presence is a hint of a noisy region (see Fig. 12.5). In this way we eschew the use of any threshold on the image intensity, thus avoiding the possible loss of faint objects.

The selection of the attributes of the structures emerging from the background has been done empirically, observing the heuristic activity of the astronomer.

The astronomer judges structures both of high intensity and of large extension as real structures, while his verdict is often doubtful if the structures are faint and small since these latter characteristics are comparable with those of spurious tracks generated by the grain patterns on the photographic plates.

In these latter cases the final verdict seems to be reached by analysing the local contrast of each candidate structure; a structure whose local contrast is high is likely to correspond to a faint object, otherwise to an uncertain track.

These considerations led us to define the following attributes in the digital environment:

Extension EXT(cs): number of pixels of a candidate structure
Local Contrast CTS(cs): (Colour(cs)−Colour(lb))/Colour(lb)
Significance S(cs): (Colour(cs)−Colour(lb))/
        Granularity(EXT(cs), Colour(lb))

where:

cs = candidate structure, *lb* = local background, i.e a set of pixels surrounding the candidate structure defined by applying the expansion morphological operator as described in Serra (1982).

Colour(cs) is the mean value of the colour of pixels belonging to *cs*.

Colour(lb) is the mean value of the colour of pixels belonging to *lb*.

Granularity is a function which evaluates the standard deviation of film grain noise as defined in Andrews and Hunt (1977) for a sampling area equal to the candidate structure extension, *EXT(cs)*, a mean intensity equal to *Colour(lb)*, and a given photographic emulsion.

This last attribute has been introduced to quantify, on the basis of the statistical model described in Andrews and Hunt (1977), the film grain noise of the image in hand.

12.3.4 An Example of Structure Evaluation

Once a structure has been described, its plausibility to be a track of an object is deduced in two steps. First the meaning of each observed property of a structure is established as if it were the only known property of the structure with reference to the real scene. This first step is performed using labelling functions (L-functions).

Secondly the meaning of all properties of a structure combined together is established, becoming an interpretation of the structure itself. This second phase can be conveniently represented in the form of Multi-Valued Logic Trees, MVLT (Garribba et al., 1985; Guagnini et al., 1989).

As far as the first point is concerned, different values of an attribute may be equivalent with respect to a specific interpretation. These values are therefore denoted by the the the same label. In general, properties are mapped into labels by means of labelling functions, where labels reflect this interpretation verdict. This partition is performed exploiting the image model.

Formally, an L-function maps a set of properties $D = (d_1, d_2, \ldots, d_n)$ into a set of labels $L = (l_1, l_2, \ldots, l_m)$, where $m < n$.

The set D is partitioned into equivalence classes. A label is associated with each class to denote the meaning of the class in the interpretation process.

The use of a whole set of defined L-functions allows the mapping of each candidate structure into a set of labels, each one relative to an attribute property, which constitute a synthetic description (s-description) of the structure itself.

The s-description is now used as input to a context evaluation function represented in the form of a MVLT (Michalski, 1980).

A MVLT can be conveniently represented by a dyadic tree (a binary tree in the terminology of Wirth, 1986) in which:

- each terminal node, called a leaf, is associated with an attribute in the s-description and therefore takes values on the set of associated labels;

EAST_WEST VALLEY
S̄C=256

RULES

(a 400) ⟶ (a 401)

(401 400) ⟶ (401 401)

(401 b) ⟶ (256 b)

(401 256) ⟶ (256 256)

a WEST

b EAST

WEST:= 1,2,3,128,129,131

EAST:= 8,16,24,32,48,56

example

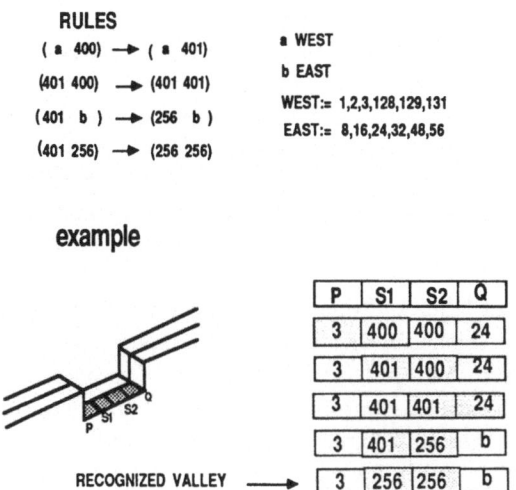

P	S1	S2	Q
3	400	400	24
3	401	400	24
3	401	401	24
3	401	256	b
3	256	256	b

RECOGNIZED VALLEY ⟶

Fig. 12.4. Example of rules used by a BL system to recognize an east-west valley. The alphabet is the set of SLOPE CODES plus the flex and minima code (400). The axiom is a CDI whose pixels have a colour value defined on the alphabet. As shown, five steps are necessary to identify an east-west valley, the length of which is four pixels.

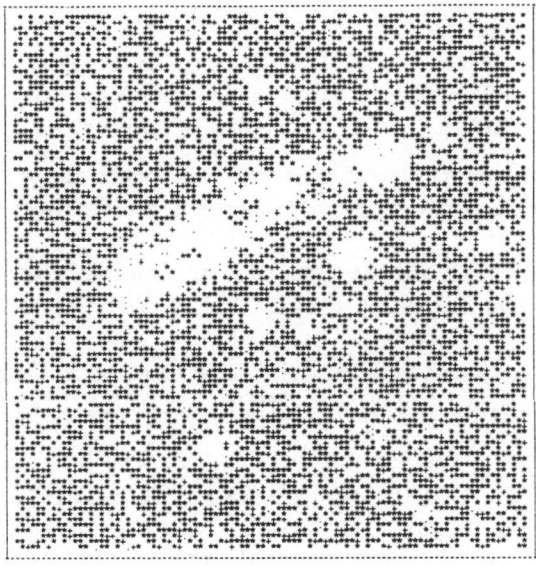

Fig. 12.5. Structures identified in the CDI shown in Fig. 12.1 (a) and characterizing the background and separating regions between close objects; '*' denotes single and multiple Ridges and Valleys, '+' Bottom valleys, '.' Gullies.

- each compound node is associated with a multivalued logic operator (op) which combines the set of labels in the leftsubtree L_1 with the set of labels in the rightsubtree L_2 to produce a new set of labels L_3 $(op : L_1 \times L_2 \rightarrow L_3)$;
- the set of labels Ltop associated with the root is the set of final classes in the interpretation process.

Thus a MVLT supplies as output the label of the class to which the structure is assigned.

This classification is performed traversing the tree structure starting from the leaves which contain the input sets of labels, computing the intermediate sets of labels by means of the multivalue operators specified in the compound nodes, until the root node is reached. A multivalue operator specified in a node is evaluated once the two sets of labels associated with the left and right subnodes become available.

The processes of label and operator definition and tree composition are empirical and subject to the uncertainties of the expert's reasoning.

In the case of structure judgement with respect to plausibility of being a real but faint track, for each attribute a set of thresholds is determined. The general definition of such thresholds exploits knowledge about:

1. The plate: plate scale (ps), seeing (see), exposition time (et), granularity (gr).
2. The digitization process: pixel size (pixs), quantization range (qr).
3. The distribution (ds) of the attribute properties.

A threshold therefore has the form: $\text{THR} = f(ps, see, et, gr, pixs, qr, ds)$.

For example, the attribute extension of a candidate structure EXT(cs) is mapped by the following L-function (LEXT) into four verdict scores:

```
LEXT= NEGLIGIBLE (N) if          EXT(cs)  < thr1
      SMALL       (S) if thr1 ≤  EXT(cs)  < thr2
      MEDIUM      (M) if thr2 ≤  EXT(cs)  < thr3
      LARGE       (L) if thr3 ≤  EXT(cs)
```

The thresholds thr1, thr2, thr3 are computed for each image exploiting knowledge about the plate (plate scale and seeing) and the digitization process (pixel size).

Thr1, thr2 and thr3 are the separators of plausible classes. A candidate structure is thus plausibly labelled as LARGE, MEDIUM, SMALL or NEGLIGIBLE.

The L-function associated with the properties of Local Contrast (LCTS) is defined so as to identify three classes (HIGH, MEDIUM, LOW):

```
LCTS = LOW    (LC) if          CTS(cs)  < Ta
       MEDIUM (MC) if Ta ≤     CTS(cs)  < Tb
       HIGH   (HC) if Tb ≤     CTS(cs)
```

The thresholds Ta and Tb are defined by examining the distribution of local contrast values for the candidate structures.

The last L-function (LS) which evaluates the properties of Significance identifies three classes (LOW, MEDIUM, HIGH) as follows:

$$
\begin{array}{lll}
\text{LS} = \text{LOW} & \text{(LS) if} & \text{CTS(cs)} < \text{T1} \\
\text{MEDIUM (MS) if} & \text{T1} \leq \text{CTS(cs)} < \text{T2} \\
\text{HIGH} & \text{(HS) if} & \text{T2} \leq \text{CTS(cs)}
\end{array}
$$

T1 and T2 are computed on the basis of the film grain noise value distribution, granularity and pixel size.

Once all L-functions have been applied to each candidate structure a verdict on their nature can be derived only by combining their associated labels by a MVLT.

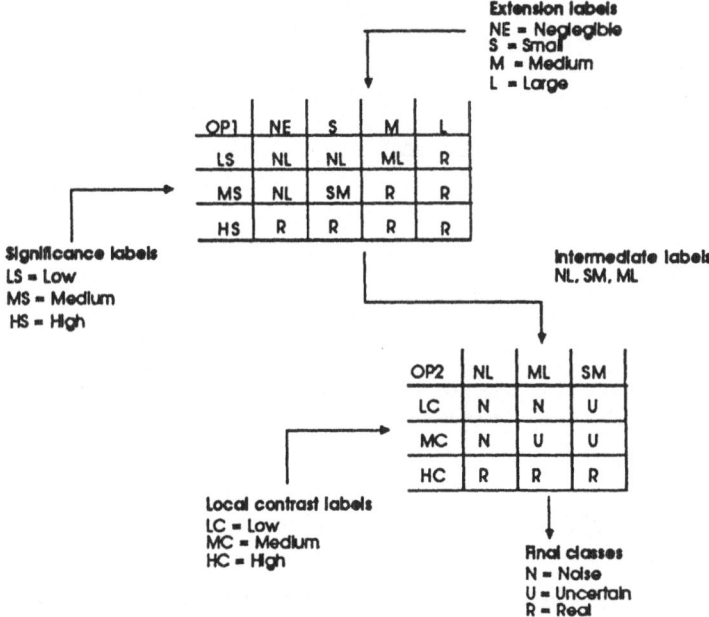

Fig. 12.6. Multi-Valued Logic Tree used to classify the candidate structures identified in the segmentation phase in Fig. 12.1 (a).

Two operators have been defined to combine the labels following the procedure adopted by the astronomer.

The operator OP1 combines the extension labels with the significance labels while OP2 combines the intermediate labels produced by OP1 with those of local contrast as shown in Fig. 12.6.

A candidate structure which is LARGE or has HIGH significance is directly classified as a track of a REAL object by OP1 even if its local contrast is LOW, since large structures or those whose signal emerges from the local background are judged as REAL objects by the astronomer. The same verdict is associated with the candidate structures which have MEDIUM size and significance. SMALL and NEGLIGIBLE tracks which have LOW or MEDIUM significance, require a further analysis in which their local contrast is taken into account. HIGH contrast is a discriminant property in order to classify the structure as

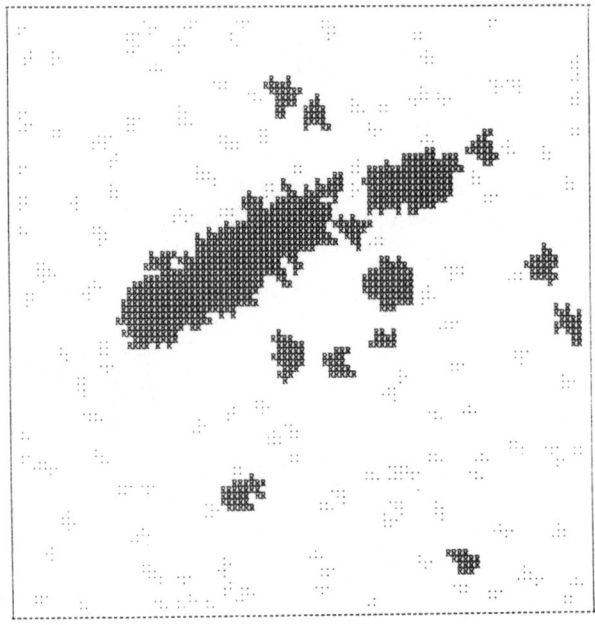

Fig. 12.7. Structure identified by the segmentation phase and classified by the MVLT in Fig. 12.6. 'R' denotes tracks classified as REAL objects, '.' tracks classified as NOISE patterns. There are no UNCERTAIN tracks in this field.

REAL, while in the other cases the structure is to be classified as NOISE or UNCERTAIN.

The final labels classify a candidate structure as REAL, i.e. a plausible track of a real source; Uncertain if the data in the image do not allow a precise verdict; or Noise track (see Fig. 12.7).

Assessment of the consistency of the proposed system may be gained through the comparison of the interpretation results of the programs and the astronomer. In our proposed approach the results of this test are summarized by a confusion matrix. In Fig. 12.8 the comparison between the result of the independent classifications of the astronomer and the programs on some digitized subfields from the ESO sky survey plate 157J are shown. These images were used to define the L-functions and the MVLT shown in Fig. 12.6 during the exploration work. The same data were used for a re-examination of the field. During this phase, each object on which a disagreement was detected was reexamined and the classification procedure discussed. This discussion improved the agreement with the astronomer up to 89%.

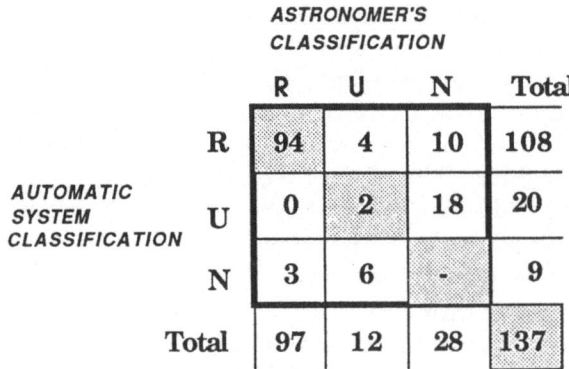

Agreement : 70.07%

Fig. 12.8. Cross comparison between the interpretation results of the astronomer and programs on sample images from ESO sky survey plate 157J by means of a confusion matrix.

12.4 Conclusions

A method for encoding astronomers' knowledge has been discussed, which allows the combined exploitation of crisp (i.e. numerical, statistically expressed) and heuristic procedures in the analysis of astronomical images.

Heuristics at the three levels of the astronomer's knowledge, the experiment model, the scene model, and the imaging process model, are translated into different types of programmed tools, according to their different natures.

The imaging process model is mainly used to define the proposition P which drives the segmentation phase, and the thresholds in the L-functions. When the structure shapes are described, or global decisions about the nature of an observed track are to be taken, also the scene and expert models play a role. For example the definition of a meaningful concavity, which can be the hint of a new feature emanating from the main body, or the definition of a noisy outlet, which cannot be considered an arm, requires considerations both on the nature of the scene (questions like: "are there any overlapping objects?" are asked) and on the expert model adopted to choose the physical parameters (object brightness, signal/noise ratio etc.) which influence the description and the decision.

Note that the imaging process, even if it is the result of a complex composition of different subprocesses, is in principle well understood, and a large agreement on the rules which are derived from it seems possible. The scene model and the experiment model appear tied to theories which are under discussion. The identification of the models influencing the rules is therefore important in order to establish their plausibility.

The formal coding of the astronomer's knowledge takes into account and exploits the context, both in shape description and judgement steps. As to the shape, CAIL systems take into account those contextual interactions among pixels which the astronomer indicates as relevant in the formation of structures. To limit the combinatorial explosion of the contextual features associated with the description of each type of structure, attributes of the symbols are exploited either to compute the composed features or to evaluate the conditions which switch on the rule firing. Both of these tools allow the reduction of the number of rules necessary to describe the set of structures of a given type and allow the user to exploit the plausible interpretation of the recognized substructures for the recognition of composed structures.

The L-functions, on their side, contextually contrast several different judgements that the astronomer can associate with a single observed property. So for example the extension of a structure (EXT) call for judgements as negligible, small, medium or large. Each of these judgements defines a fuzzy set on EXT (Di Gesù, 1987). The L-function LEXT maps directly a property into the most plausible judgement. If the astronomer changes one (or more) of his judgement possibility functions, all the values must be consequently remapped.

On the whole, the proposed approach faces the combinatorial explosion by adopting a predicative form to express rules, L-functions and MVLTs, rather than a propositional form, as is usual in similar cases. Each attributed conditional rule summarizes a large set of traditional non-attributed, non-conditional antecedent-consequent couples, and also L-functions and MVLTs summarize sets of traditional IF-THEN rules (Fu, 1982; Mussio et al., 1989). Here again the three models are exploited to define both the attributes and the conditions which are meaningful in the experiment as well as to develop judgement devices when property values, which are equivalent with respect to the judgement goals, are grouped. Once the proposed approximate description of the astronomer's knowledge is codified into programs it allows the repetition of experiments whose results are systematically comparable and whose reliability can be evaluated by the knowledge of sample confusion matrices.

The proposed approach has some pros and cons. From the astronomer's point of view, it allows the definition of a more synthetic, possibly more expressive definition of the generative and judgment devices. In this way both the discussion and criticism of the rules, L-functions and MVLTs as formal tools for the detection of celestial object tracks become easier. When their programmed version is used to illuminate their actual meaning on new data, it becomes easier to follow the tracks of the performed computation. In case of disagreement between the astronomer and the computed results, this synthesis leads to a faster understanding of the points in which the two analyses differ. As a consequence, one can more quickly understand the meaning of the tools, the improvement and enrichment of knowledge stored in the programs both in quality and in quantity. From the formal side, however, these tools are not yet well established. The syntactical correctness, completeness and consistency of each proposed tool can be only proven at the empirical level, by practical experiments on data, because the theories of CAIL systems, L-functions and MVLTs are still under development. Furthermore, the rules, L-functions, and MVLT are at present derived from the experience of only those astronomers who participate in the system design. On the other hand, once this knowledge is coded into synthetic formal tools, it can be studied and criticized by different researchers. Criticism can be discussed at large and translated into new updated versions of programs and tested on old and new data. Once accepted, these new rules constitute an enlargement and improvement of the knowledge stored in the system.

References

1. Accomazzi, A., Bordogna, G., Mussio, P. and Rampini, A. (1989), "Rule-based description and plausible classification of objects in digitized astronomical images", *Data Analysis in Astronomy III*, V. Di Gesù, L. Scarsi, P. Crane, J.H. Friedman, S. Levialdi and M.C. Maccarone (eds.), Plenum Press, New York, in press.

2. Ahuja, N. and Shachter, B.J. (1981), "Image models", *ACM Computing Surveys*, **13**, 363–398.

3. Andrews, H.C. and Hunt, B.R. (1977), *Digital Image Restoration*, Prentice-Hall, New Jersey.

4. Balestreri, M., Della Ventura, A., Fresta, G. and Mussio, P. (1979), "A proposed pictorial language for describing astronomical objects", *Image processing in Astronomy*, G. Sedmak, M. Capaccioli and R.J. Allen (eds.), Osservatorio Astronomico di Trieste, Italy, 268–298.

5. Della Ventura, A., Rampini, A., Rabagliati, R. and Serandrei Barbero, R. (1987), "Development of a satellite remote sensing technique for the study of alpine glaciers", *International Journal of Remote Sensing*, 8, 203–215.

6. Di Gesù, V. (1987), "Fuzzy sets and data analysis", *Astronomy from Large Data Bases: Scientific Objectives and Methodological Approaches*, F. Murtagh and A. Heck (eds.), European Southern Observatory, Garching bei München, 183–197.

7. Fu, K.-S. (1982), *Syntactic Pattern Recognition and Applications*, Prentice Hall, New Jersey.

8. Garribba, S., Guagnini, E. and Mussio, P. (1985), "Multiple-Valued Logical Trees: meaning and prime implicants", *IEEE Transactions on Reliability*, **R-34**, 463–469.

9. Godwin, J.G., Metcalfe, N. and J.V. Peach (1983), "The Coma cluster-I", *Monthly Notices of the Royal Astronomical Society*, **202**, 113–124.

10. Guagnini E., Mari L. and Mussio P. (1989), "Valutazione plausibile dei comportamenti di sistemi dinamici da loro descrizioni sincroniche", *Dinamica dei Sistemi*, in press.

11. Hubble, E. (1936), *The Realm of the Nebulae*, Yale University Plenum Press, New Haven, CT.

12. King, I.R. and Raff, M.I. (1977) *Publications of the Astronomical Society of the Pacific*, **89**, 120–121.
13. Kurtz, M.J. (1989), "Classification and knowledge", this volume.
14. Merelli, D., Mussio, P. and Padula, M. (1985), "An approach to the definition, description, and extraction of structures in binary digital images", *Computer Vision, Graphics and Image Processing*, **31**, 19–49.
15. Michalski, R.S. (1980), "Pattern recognition as rule-guided inductive inference", *IEEE Transactions on Pattern Analysis and Machine Intelligence*, **PAMI-2**, 349–361.
16. Murtagh, F. and Lauberts, A. (1986), "A curve matching problem in astronomy", *Pattern Recognition Letters*, **4**, 456–469.
17. Mussio, P. (1985), "Design of a pattern-directed system for the interpretation of digitized astronomical images", *Data Analysis in Astronomy*, V. Di Gesù, L. Scarsi, P. Crane, J.H. Friedman and S. Levialdi (eds.), Plenum Press, New York, 227–234.
18. Mussio, P., Padula, M. and Protti, M. (1989), "Attributed Conditional L-systems. A tool for image description", *Proceedings of the 9th International Conference on Pattern Recognition*, Rome, in press.
19. Pólya, G. (1945), *How To Solve It*, Princeton University Press, New Jersey.
20. Pradé, H. (1985), "A computational approach to approximate and plausible reasoning with application to expert systems", *IEEE Transactions on Pattern Analysis and Machine Intelligence*, **PAMI-7**, 260–283.
21. Salomaa, A. (1973), *Formal Languages*, Academic Press, New York.
22. Sandage, A. and Sandage, M. (1983), *Galaxies and the Universe*, University of Chicago Press, Chicago.
23. Serra, J. (1982), *Image Analysis and Mathematical Morphology*, Academic Press, New York.
24. Thonnat, M. (1985), "Automatic morphological description of galaxies and Classification by an expert system", Research Report, INRIA Sophia Antipolis, Valbonne, France.
25. Wirth, N. (1986), *Algorithms and Data Structures*, Prentice-Hall International Editions, New Jersey.

New Directions

13 Connectionism and Neural Networks

H.-M. Adorf
Space Telescope — European Coordinating Facility
European Southern Observatory
Karl-Schwarzschild-Str. 2
D-8046 Garching bei München
F.R. Germany

13.1 Part I: Background

13.1.1 Introduction

It has long been known that, on a microscopic level, the mammalian brain is inherently slow. Its operational speed is estimated to be only between 100 and 1000 Hz. Yet, it is capable of recognizing a visual scene and issuing a reaction within fractions of a second, i.e. within about 100 cycles. Thus on complex cognitive tasks the brain easily outperforms the central processor unit of even the most powerful contemporary serial computer, which operates at frequencies of tens of MHz.

To many people the solution to this puzzle seems to be "massive parallelism": the brain contains a huge number (something between 10^{10} and 10^{11}) of rather simple computing elements, called "neurons", which are highly interconnected (about 10^3 to 10^5 connections per neuron) through "axons" and "dendrites". Though its individual components are inherently slow, the system as a whole operates *fast*, since many "computations" are carried out in parallel. In addition the brain has the ability to *learn*, a property which distinguishes it sharply from an ordinary computer system, which has to be *programmed* to perform a meaningful task.

In the past decades biologists and neurophysiologists have greatly improved their understanding of the organizational principles of brains. In the picture which has emerged, brains are composed of a network of neurons, i.e. cells that can amplify and conduct electrical pulses. From the main body of each neuron a long fiber, the *axon*, emanates branching into a number of *dendrites*, which end on or near the bodies of other neurons. The coupling (or *synaptic junction*) between the dendrite and the next cell body may be such that an arriving nerve pulse has an *excitatory* or an *inhibitory* effect on the recipient neuron.

At the present time computer scientists are busy capturing the essentials of biological parallel models of "computation" and embodying them in *artificial neural networks* (ANNs), sometimes also called *artificial neural systems* (ANSs), *parallel distributed processing* (PDP) or *connectionist models*.

Generally speaking, an *artificial neural network* is a dynamical, information processing system composed of a large number of simple *processing elements* (PEs), also called *artificial neurons* (Fig. 13.1), which are interconnected by directed links, called *interconnects* or *connections* and which co-operate to solve a computational task. Each PE locally combines inputs from other PEs and computes a single output value. In the simplest case it forms a weighted sum of all inputs and compares that with a threshold. If the net input exceeds the threshold, the neuron output is set to "high-value", otherwise to "low-value".

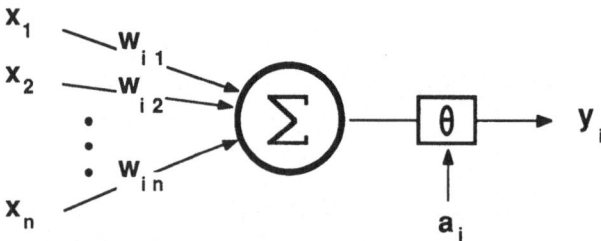

Fig. 13.1. Artificial neuron after McCulloch and Pitts. Inputs $x_1 \ldots x_n$ coming e.g. from other neurons are multiplied with weights $w_{i1} \ldots w_{in}$, summed up and finally thresholded using bias a_i. The threshold (step) function may be replaced by a more general transfer function.

Depending on which aspects of their complex biological counterparts are emphasized, different network models result, which are characterized by differences in their *neuron behaviour*, *network topology* and *learning rules*. In the course of time, the artificial neural network paradigm has become a rich and rewarding source of inspiration; networks are being studied by computer scientists as attractive computational models with interesting information-processing properties, irrespective of biological fidelity.

The concept of neural networks is expected to be beneficial beyond the high computation rates provided by massive parallelism. Neural nets are fault tolerant: the failure of a single neuron does not normally cause the whole system to break down. Many neural nets incorporate an *adaptation mechanism*, by which they can react to their "environment" and improve performance in the course of time. Some neural networks have convincingly developed the ability of *abstraction* and *generalization*, tasks which have withstood the attempts of many classically oriented AI researchers to emulate.

The architecture of a neural network represents a radical departure from that of a conventional serial computer, the design principles of which were conceived in the 1940s by the famous Hungarian mathematician John von Neumann. The von Neumann architecture is characterized (Rosati et al., 1987) by a clear separation of the central processor unit and its memory, with a communication link in between. In the neural net approach, this distinction dissolves. The memory *is* the computer. Information, which in the conventional architecture can be clearly located in identifiable storage cells, is stored in a distributed manner as weights of the network connections.

A few years ago a comprehensive review of all interesting and important developments in the field of artificial neural networks might perhaps have been feasible. In view of the recent explosive growth in the number of relevant publications, an appraisal of the entire spectrum has become almost impossible and my review will therefore be based on a subjective selection. After a brief historical sketch I will highlight some of the aspects which make artificial neural networks worthy of consideration within astronomy.

13.1.2 Historical Development

13.1.2.1 The Pioneering Years. The landmark paper of Warren McCulloch and Walter Pitts (1943) is often taken as the starting point of ANN research. McCulloch and Pitts considered networks of two-state threshold elements and proved that every logical function could be implemented using these kinds of "neurons". Their result implies that any finite-state machine can be simulated by a network of such neurons (cf. Arbib, 1987, p. viii).

Hebb's *The Organization of Behaviour*, published in 1949, was paving ground for many neural network models of learning and adaptation. Hebb proposed a learning scheme for formal neurons like those of McCulloch and Pitts in which connections between neurons were strengthened whenever the neurons "fired" jointly.

Enthusiasm for neural networks peaked for the first time when in the late 1950s Frank Rosenblatt and his colleagues at Cornell University invented the *Perceptron*, a one-layer neural network which could be trained to optimally recognize "linearly separable" categories. Networks with multiple layers, however, were poorly understood at that time. There was no theory for multilayer training and it was not even known whether multilayer networks would offer anything beyond the capabilities of single-layer networks.

Funding of an integrated study of natural and artificial intelligence seized up in the mid-1960s, after Marvin Minsky and Seymour Papert, two pioneers of artificial intelligence, had visited the US Pentagon's Advanced Research Projects Agency (now DARPA) where they convincingly demonstrated that the Perceptron was incapable of solving a simple, yet important, classification problem known as the "exclusive-or" problem (Rosenfeld, 1987).

13.1.2.2 The Resurgence of the Field. Though reduced in intensity, research on
neural networks was not abandoned totally. A few researchers in the areas of neu-
rophysiology, biological control theory, cognitive psychology and artificial intelligence persistently carried on in their research, unaffected by the lack of interest
and support from the rest of the world, notably Shun-ichi Amari, James Anderson, Michael Arbib, Kunihiko Fukushima, Stephen Grossberg, Teuvo Kohonen,
Arthur Little, Christoph von der Malsburg and Paul Werbos (see Box 13.1).
These people experienced difficulties to get their work published and, as a consequence, it is scattered among a variety of journals.

Box 13.1: Highlights in the development of artificial neural networks

1943	W. McCulloch & W. Pitts	threshold logic neuron
1949	D. Hebb	neuron learning rule
1955	O.G. Selfridge	Pandemonium pattern recognition model
1957	A.N. Kolmogorov	function representation theorem
1959,62	F. Rosenblatt	Perceptron
1960	B. Widrow & M.E. Hoff	Widrow-Hoff learning rule, ADALINE
1961	K. Steinbuch	Lernmatrix
1966	J. von Neumann	cellular automata
1968	S. Grossberg	instar, outstar, avalanche, ART
1969	M. Minsky & S. Papert	theory of Perceptrons
1970	J.A. Anderson	associative memory
1973	J.A. Anderson	Brain-State-in-the-Box (BSB)
1973	C. von der Malsburg	self-organization
1974	T. Kohonen	associative memory
1974	P. Werbos	error back-propagation (PhD thesis)
1980	K. Fukushima	Neocognitron
1982	J.J. Hopfield	stochastic binary Hopfield net
1982	T. Kohonen	Kohonen feature map
1983	S. Kirkpatrick, C. Gelatt & M. Vecchi	simulated annealing
1984	J.J. Hopfield	deterministic "graded" Hopfield net
1985	Y. Le Cun	learning in asymmetric nets
1985	D. Parker	error back-propagation (rediscovery)
1985	D.H. Ackley, G. Hinton & T. Sejnowski	Boltzmann machine
1986	D. Rumelhart, G.E. Hinton & E.J. Williams	multilayer Perceptron
1986	H. Szu	Cauchy machine
1987	G. Carpenter & S. Grossberg	Adaptive Resonance Theory (ART) II
1987	R. Hecht-Nielsen	generality of three-layer network

In the course of time, training algorithms were developed for multilayer networks (Werbos, 1974; Le Cun, 1985; Parker, 1985; Rumelhart et al., 1986),
which could solve the "exclusive-or" problem. This led to the recognition that
networks with multiple layers are indeed more powerful than single-layered ones.
In the meantime computer processing power increased dramatically, allowing the
simulation of quite large networks on serial processors.

A couple of other developments have raised interest in ANNs to the high
level we see today:

• The generality and sufficiency of three-layer feed-forward ANNs for function mapping has been recognized (Kolmogorov, 1957; Hecht-Nielsen, 1987).

• The dynamics of restricted classes of ANNs — particularly those with symmetric connections — has been theoretically analyzed and largely understood (see e.g. Hopfield, 1982, 1984; Amit et al., 1985a, b; Feldman, 1985; Goles and Vichiniac, 1986; Hirsch, 1987).

• Physicists have become interested in "spin glasses" described by an interaction model which is mathematically equivalent to a Hopfield neural network (see e.g. Grötschel et al., 1986; Hertz et al., 1986; Mézard et al., 1986).

• Stochastic optimization methods such as *simulated annealing* (Kirkpatrick et al., 1983; Geman and Geman, 1984; Szu, 1986; Aarts and van Laarhoven, 1987; Davis, 1987) have been (re-) discovered. Combined with the neurocomputing paradigm they have led to network architectures suitable for combinatorial optimization (Ackley et al., 1985; Baum, 1986; Jeffrey and Rosner, 1986; Bounds, 1987; Bridle, 1987; Feldman, 1987; Hinton and Sejnowski, 1987; Johnson, 1987; Ramanujam and Sadayappan, 1988).

• There is growing recognition of the importance of *parallel architectures* and *algorithms* (see e.g. Kumar and Kanal, 1984; Karp and Widgerson, 1985; Kibler and Conery, 1985; Lapedes and Farber, 1986; Tanimoto, 1986; Blelloch and Rosenberg, 1987; Hillis and Barnes, 1987; Hwang, 1987; van Biema, 1987; Waltz, 1987; and Genthner et al., 1988).

• Advances in VLSI technology have opened the possibility of hardware implementations of neural network circuits (see e.g. Graf et al., 1986, 1988; Sivilotti et al., 1986; Thakoor et al., 1987).

It seems that the current surge of interest in neural network was largely stimulated by two articles written in 1982 and 1984 by John Hopfield, professor of biology and chemistry at Caltech. It has been estimated (Rosati et al., 1987) that Hopfield has brought 80 % of the current full-time neural network researchers to the field.

13.1.3 Concepts

13.1.3.1 Mathematical Description. In algebraic notation a model for a feedback network of McCulloch-Pitts threshold neurons would read as follows: let $\mathbf{y} = (y_1, \ldots, y_n)^T$ be the output of neurons $1, \ldots, n$ and W_{ij} denote the connection weight corresponding to output y_j and neuron i. Then neuron i receives an input contribution of $W_{ij} y_j$ from neuron j. All contributions add up to the net input of $\sum_j W_{ij} y_j$. If a_i denotes an external bias of neuron i, then the output of neuron i is given by $y_i' \leftarrow \Theta(\sum_j W_{ij} y_j + a_i)$, where Θ denotes the Heavyside step function: $\Theta(x) = 1$ if $x > 0$, and 0 otherwise. The threshold function above may be replaced by a more general "transfer function" σ_i, which is usually required to be strictly monotonic and to map values from the interval $(-\infty, +\infty)$ into $(-1, +1)$ or $(0, +1)$. If the transfer function is monotonic and "S-shaped", it is referred to as a *sigmoid* function.

13.1.3.2 Network Topology. The functional ability of a neural network is largely dependent on its node topology, i.e. the number and arrangement of interconnects between the neurons. Often neurons can be naturally grouped into layers, depending on their functionality within the network.

A standard network topology, quite often used for pattern recognition and classification tasks, is the *feed-forward* network with three neuron layers (Fig. 13.2). The first layer, the *input layer*, receives the values of features describing an input pattern, e.g. the intensity values of an image. Each neuron of the *output layer* corresponds to a separate class. Neurons in an *intermediate* or *hidden layer* provide the necessary mapping between the input and output neurons in such a way that out of all output neurons only that neuron eventually fires which gives the correct classification of the input pattern.

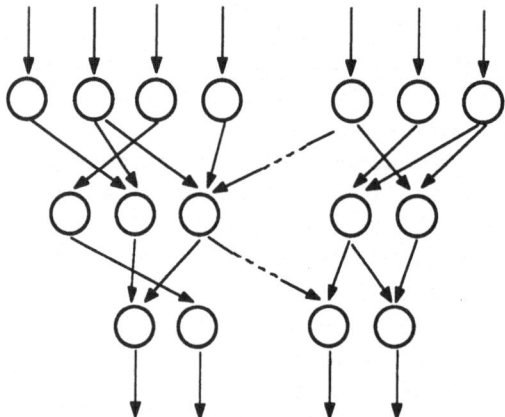

Fig. 13.2. Feed-forward network with three neuron layers. This network topology is frequently used for pattern recognition and supervised classification. Input signals, e.g. sensory data, stimulate the neurons of the top layer. Their output is transformed in the middle layer. The bottom layer produces the recognition/classification result.

The most general network topology is characterized by complete feedback, i.e., the output line of each neuron is (re-) used as input to the network (Fig. 13.3). All other network topologies are special cases of the complete network with some weights simply being set to zero.

13.1.3.3 State Changes — Fast Network Dynamics. Feed-forward networks are very fast: the delay between the presentation of the input pattern and the resulting classification is just the time constant of a single network layer times the number of layers. Thus real-time applications are within the range of such network topologies.

In general, feedback networks show a more complicated dynamical behaviour. The state of the network is completely described by its output vector which can be thought of as a point in an n-dimensional state space. A given vector used as input to a neural network will generate a transformed vector, which will usually be different from the input. If both vectors incidentally agree with each other, they represent a fixed point of the system. Fixed points of a neural network correspond to the stored memories.

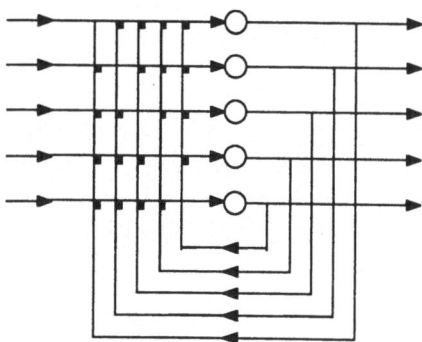

Fig. 13.3. Hopfield network. Its topology is characterized by complete feedback (excluding, however, auto-feedback). All other network topologies are special cases of this one.

The dynamics of feedback networks with *symmetric* connections between the neurons is conveniently analyzed and understood with the help of a so called Lyapunov or *energy* function (see e.g. Willems, 1970), which can be derived from the connection weights and biases (Hopfield, 1982, 1984; Goles and Vichiniac 1986). The important property of this function is that it does not increase under the usual network dynamics. The stable fixed points of the network dynamics exactly correspond to the local minima of the network energy.

13.1.3.4 Learning — Slow Network Dynamics. Another kind of dynamics, which occurs on quite a different time scale, involves the change of weights, a process which is called *learning*. Learning is used to change the network structure in order to adapt the network to static or slowly changing environmental conditions.

The ability of neural networks to learn is one of the main reasons why they are so extensively studied at the moment. Two different mechanisms can be distinguished here: in *supervised* learning the network adjusts connection weights in response to labeled examples from a training set; in *unsupervised* learning the network organizes its internal structure without the help of a "teacher". Both kinds of learning occur in biological systems, but it is not exactly known how learning is controlled there. In artificial systems learning rules are often designed such that a given objective function is optimized. For a review of network learning procedures see Hinton (1987).

13.1.3.5 Summary. In summary, ANNs differ in the kind of neurons from which they are made, the way neurons are interconnected and the type of updating rule governing the network dynamics. Thus, homogeneous ANNs can be classified along the following dimensions:

1. individual neuron behaviour:
   a. the *output range* of individual neurons (identical to the range of the neuron transfer function) may have two values, have finitely many discrete values, have countably infinite values or be a continuous set;
   b. the *type* of the individual neuron's transfer functions may be deterministic or stochastic.
2. connections:
   a. the *network topology* may be feed-forward only or include some feedback;
   b. the *connection matrix* may be symmetric or asymmetric;
   c. *connection weights* may be fixed (no learning) or modifiable (learning); in the latter case different learning algorithms may be used to adjust the weights.
3. dynamics:
   a. the *time axis* along which the network dynamics evolves, may be continuous or discrete;
   b. the network *update rule* may be synchronous (all neurons are updated simultaneously) or asynchronous (neurons are updated one after the other); in the latter case *neuron selection* may be deterministic or stochastic.

Homogeneous networks may be coupled together to form multiple heterogeneous networks allowing additional degrees of freedom, such as different time constants for subnets.

13.1.4 Hard- and Software for ANNs

Although interest in artificial neural networks is, to some extent, spurred by the prospect of enormous speed-ups to be gained from implementations on specialized hardware, at the present time most ANN research is still being carried out using software simulators on conventional, general-purpose serial processors.

13.1.4.1 Software Simulators. Neural network simulators are flexible in terms of modifiability. They are necessary for the initial stages in the development of new network models and algorithms, but on sizable problems may suffer from performance problems.

A couple of non-commercial simulators of this kind have been developed, mainly at American universities and are often available to academia for little or no cost. Some of these are *P3*, *SFINX*, *RCS*, *JANET*, *SPREAD-3*, and *ANNE* (see Box 13.2).

On the commercial side things have evolved quickly; while in 1986 only one neural modelling tool was available, there are now at least a dozen. The more notable ones are the *ANZA Basic Netware Package* from Hecht-Nielsen Neurocomputer Corporation, the *ANSkit* and *ANSim* from Science Applications International Corporation, the *Nestor Development System* from Nestor, Inc., (*AI Magazine 9*, No. 2, p. 146), the *N-NET* toolkit from AI Ware, Inc., the *Neural Network Development Tools* from NeuralWare, Inc., and *NeuroShell* from Ward Systems Group (*AI Expert*, Oct. 1988, p. 67). These tools are developed to aid rapid prototyping and concept testing of neural networks. For a review of commercially available ANN software simulators see Schwartz (1988).

Box 13.2: Non-commercial network simulators

The *P3* system (Zipser & Rabin 1987) was developed at the University of California to support the research of the PDP group. It is written in Lisp.

The *SFINX* neural network simulation environment (Paik et al., 1987) was developed at the Machine Perception Laboratory of the University of California at Los Angeles. It has been designed to flexibly support interactive investigations of various neural structures, in particular large, highly regular networks. The simulator is written in C and YACC and should port to most machines under UNIX. SFINX has been used to simulate ANNs for image segmentation, shape recognition and other low-level vision tasks.

The *Rochester Connectionist Simulator* was developed at the University of Rochester (Goddard et al., 1987; Diederich, 1988) out of the Lisp-based *ISCON* system. RCS is written in C and runs on general purpose UNIX workstations (in particular on SUNs). Neural network functionality is provided through calls to library functions, which include standard models like backpropagation and reinforcement learning. RCS possesses an elaborate graphics interface. It is available for a nominal fee of 150 $ (US).

JANET is essentially a re-implementation of RCS in Common Lisp and runs on, e.g., Symbolics and MacIntosh workstations. JANET's enhanced graphics especially support "thermodynamic" PDP models. For non-commercial organizations JANET is available cost-free on an "as is" basis directly from the German Gesellschaft für Mathematik und Datenverarbeitung (GMD), Bonn (Hernandéz 1988).

SPREAD-3 (Diederich & Lischka, 1987) was developed by the GMD. It is implemented in Lisp on Symbolics workstations.

The *ANNE* neural network simulation system was developed at the Oregon Graduate Center for Intel's multiprocessor iPSC machine.

13.1.4.2 Hardware Implementations. The possibility of physically casting neurons and their interconnects directly into VLSI hardware is being investigated by several research teams (see Graf et al., 1986, 1988; Hwang, 1987; Hwang et al., 1987; and Hutchinson et al., 1988). Speed-up factors of up to 10^6 have been reported for specialized chips compared to software emulation on a standard VAX. The first book on VLSI designs for ANNs has already appeared.

Implementations in electronic circuits run into difficulties when large networks with high connectivity are required, but perhaps developments in the optical computer arena may overcome these problems (see Farhat, 1987, in the *Applied Optics* special issue on neural networks). Ultimately networks based on neuro-hardware should achieve very high performance, but the whole area has still to mature before we will be able to buy neuro-chips and -boards in the same way as we can buy memory-boards these days.

13.1.4.3 Co-Processors. In between these two extremes are specialized ANN co-processors, devised to be plugged into general purpose workstations or PCs. These accelerator boards can be programmed to realize any desired neuron transfer function, network topology and update rule. Function libraries are provided with the boards allowing some neural network functionality to be invoked from within a conventional program. The best known examples in this area are the *ANZA* neurocomputing coprocessor system board from Hecht-Nielsen Neurocomputer Corporation, the Δ-1 floating point processors and the Σ-0 and Σ-1 neurocomputer workstations from Science Applications International Corporation, and TRW's *Mark III* and *Mark IV*.

13.2 Part II: Applications

Artificial neural network approaches have been suggested and used for a variety of practical tasks with promising results, notably image processing (Szu and Messner, 1986; Simpson, 1987; Simpson and Reinhard, 1988), natural language processing (Fanty, 1986), pattern and speech recognition (Mueller and Lazarro, 1986; Bridle, 1987; Wechsler, 1987; Watrous and Shastri, 1987; Fukushima, 1988a, b; Kohonen, 1988; Widrow and Winter, 1988), speech synthesis (Sejnowski and Rosenberg, 1986; Sejnowski, 1987), error correcting code (Platt and Hopfield, 1986), expert systems (Touretzky and Hinton, 1986; Gallant, 1988), adaptive control (Ritter and Schulten, 1987; Shapiro, 1988; Ritter et al., 1989), job shop scheduling (Foo and Takefuji, 1988a, b), supervised and unsupervised classification (Eklundh et al., 1986; Kohonen, 1987a, b; 1988; Carpenter and Grossberg, 1988; Linsker, 1988), constraint satisfaction problems (Tagliarini, 1987) and combinatorial optimization (Ackley et al., 1985; Hopfield and Tank, 1985; Baum, 1986; Jeffrey and Rosner, 1986; Tank and Hopfield, 1987; Bounds, 1987; Bridle, 1987; Feldman, 1987; Hinton and Sejnowski, 1987; Johnson, 1987). For reviews see Anderson (1986) and Lippman (1987).

In the following I will concentrate on some neural network related developments which have been carried out within the "ST-ECF Artificial Intelligence Pilot Project" (Adorf and Johnston, 1987). I will first introduce an instructive example, the N-queens problem, which will illustrate many important concepts of neurocomputing. In fact, it was our early success with a network based approach to the N-queens problem which triggered the development of network algorithms for two areas of practical interest in astronomy, namely classification and telescope scheduling.

13.2.1 The N-Queens Problem

The N-queens problem was first formulated in the 19th century for the case $N = 8$ (Campbell, 1977). The problem consists of placing N chess-queens on an $N \times N$ board such that no queen threatens another one. Phrased differently, N pieces shall be placed on an $N \times N$ board such that no row, column or diagonal holds more than one piece.

The N-queens problem is a special instance of a general class of problems known as *constraint satisfaction* or *consistent labelling* problems, which are of general interest in computer science. The N-queens problem is very suitable for studying the performance of search algorithms, since solutions exist for any $N \geq 1$, except $N = 2, 3$ (Hoffmann et al., 1969); thus the problem can be scaled to any desired size. It has long been known that the classical 8-queens problem has exactly 92 solutions and the number of solutions has been computed for all N up to 13, but beyond this, little is known about the N-queens problem. It seems that the number of solutions grows exponentially with N, and that the search effort to find a single solution using a standard deterministic backtrack algorithm also grows exponentially (Stone and Stone, 1987).

13.2.1.1 Representing the N-Queens Problem in a Neural Network. It is rather straightforward to translate the specific constraints of the N-queens problem into connections and bias terms of a Hopfield neural network. Let each square of the board be identified with a neuron and let a queen on a square be represented by the corresponding neuron being in its "on"-state. Conversely, an empty square is represented by a neuron in its "off"-state. The constraints of the problem are represented in the connections of the network as follows: each potential threat translates into an inhibitory link between the corresponding two neurons.

In the following I will concentrate on the case $N = 8$, but almost everything being said will readily carry over to the general case. Let us first look at the connections in some more detail: a queen placed on square $A1$ say, will threaten all squares in the first row, in column A and in the diagonal running from $A1$ to $H8$. So inhibitory links will have to run from $A1$ to all of those squares (Fig. 13.4). Conversely, a queen placed on any of these squares would threaten the square $A1$, so an inhibitory link should start in each of these squares and end in $A1$. This symmetry property of the connections is very welcome, since it implies a symmetric connection matrix and thus the existence of a well-behaved energy function.

The inhibitory connections discussed so far take care only of the "no threat" constraints between queens. The condition that *some* queens should be placed on the board can be taken into account by applying an excitatory bias to all the neurons.

13.2.1.2 Network Dynamics. The previous considerations fix the network topology, but do not determine the network dynamics necessary to find "fixed points" of the system, which represent solutions to the problem. Historically we first implemented a synchronous, deterministic updating rule to be used in conjunction with a continuous Hopfield-net. In this case neuron outputs take values from a continuous range and the network dynamics is described by a differential equation (Hopfield, 1984).

If the network described so far is started with an empty board (all neurons switched off), nothing happens, since the system state is already in an (unstable) fixed point of the Hopfield dynamics. This failure is caused by the fact that the N-queens problem is too symmetric. Each queen is competing with the others (on equivalent squares) in her struggle for existence. The problem can be cured by "breaking the symmetry" in either of two ways: (i) adding a small random bias to each of the neurons or (ii) placing a few queens non-symmetrically on the board. In the latter case the system actually works as a *distributed associative memory* (DAM) or *content addressable memory* (CAM): it recalls a "solution"

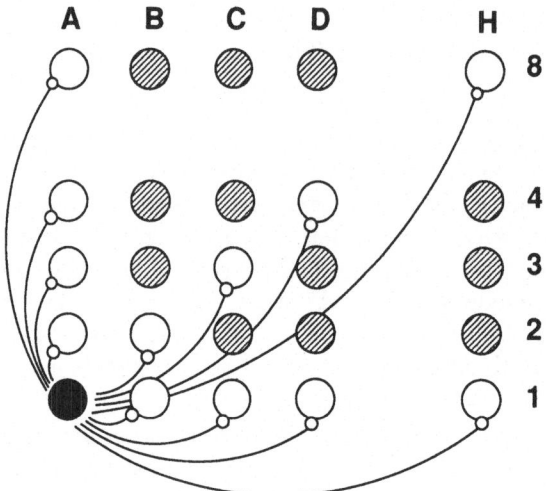

Fig. 13.4. Neural network for the 8-queens problem. Each square on the chessboard is identified with a neuron. Inhibitory connections emanating from neuron A1 are shown. If this neuron (filled circle) is activated, it inhibits all neurons in column A, row 1 and diagonal A1-H8 (white circles).

from its memory (the connections) using the set of predefined queen-positions as a "key".

It is quite instructive to watch a screen display showing the neural activations and observe how the neurons co-operate and compete while the net is working on the problem. With continuous neurons, queens grow into "partial" existence and, being in competition with others, may survive or loose their lives. After a while the network usually settles down into a state with a few neurons being completely turned on while the rest are completely turned off.

Unfortunately, it too often happens that less than 8 queens establish themselves on the board. The reason for this peculiar behaviour is that the implicit condition that exactly 8 queens should be placed on the board has nowhere been taken into account. Actually the network described above solves a slightly different problem than the one we intended: it places non-attacking queens on a chessboard such that they together "dominate" the board, i.e. each square is threatened by at least one queen. On an 8 × 8 board the minimum number of dominating queens is 5; the maximum, of course, is 8.

Viewed in energy space, a "solution" with less than 8 queens corresponds to a *local* energy-minimum, while a solution to the original 8-queens problem corresponds to a *global* minimum. The network dynamics performs a descent in energy space until the network state gets finally trapped in a local (rather than a global) energy minimum.

Another problem with the continuous Hopfield-net model with deterministic neuron updating is that it is very slow, when simulated on a serial digital computer.

13.2.1.3 Global Combinatorial Optimization — A Tough Problem. The difficulty encountered with the 8-queens problem is very typical in the global optimization area, which is known to be computationally demanding. In fact most of the interesting combinatorial optimization problems are "NP-hard". For these problems no "efficient" algorithms are known that are guaranteed to find an optimal solution in polynomially bounded time; the only algorithms discovered so far are of exponential time complexity, which means that the execution time of the algorithms scale exponentially with the problem size.

Computationally, NP-hard problems are at least as hard as problems from another class known as "NP-complete" in the computer science literature. All NP-complete problems are equivalent in the sense that they can be transformed in polynomial time into each other. Thus if an efficient (i.e. polynomially time-bounded) algorithm exists for one of these problems, it will also solve all the others in polynomial time. Currently it is not known — and indeed it is one of the most challenging open problems in computer science — whether polynomially bounded algorithms for NP-complete problems exist, but have not yet been discovered, or whether they cannot exist. (For overviews of computational complexity theory see Garey and Johnson, 1979; Cook, 1983; Karp, 1986)

Coming back to optimization problems, the objective has to be lowered: instead of optimal solutions one often has to be content with "good" suboptimal solutions. In the neural network arena the most popular approximative method for global optimization problems has been a stochastic technique known as *simulated annealing* (Kirkpatrick et al., 1983; Aarts and van Laarhoven, 1987). The combination of a discrete Hopfield net with simulated annealing is termed a *Boltzmann machine*. In a system treated with simulated annealing, states have non-zero transition probabilities to states of higher energy. Thus the system is able to wander around a stable fixed point and eventually escape from its basin of attraction. It can be proven that with probability one a globally optimal (minimum energy) state will eventually be reached, but this process may take a considerable amount of time.

For continuous systems a different method has been suggested by Jeffrey and Rosner (1986). They follow the trajectory defined by the differential equation which governs the state evolution, until the system point gets trapped in a local energy minimum. Then for some time steps the energy surface is "inverted", i.e. minima are transformed into maxima and vice versa, giving the system point a chance to escape from the local optimum. Some time later the surface is inverted again and the system is (hopefully) attracted by another local minimum. Keeping a record of all minima traversed so far allows a "good" solution to be picked for the optimization problem. Nothing, however, guarantees that a global optimum will be hit upon, but Jeffrey and Rosner report encouraging results from this *ad hoc* approach.

13.2.1.4 The Guarded Discrete Stochastic Network. In our collaborative effort at the Space Telescope — European Coordinating Facility (ST-ECF) and the Space Telescope Science Institute (STScI) we have followed yet another route to the global optimization problem. We have used a network of binary neurons together with a heuristic, controlled stochastic neuron update rule. Some special "guard" neurons are charged with the task of enforcing the global optimality condition. Each of these guards "watches" a special set of neurons (normally a row) on the board and is activated if the supervised neuron set does not contain at least one "on"-neuron. Guard neurons are coupled to the main network in an asymmetric way and effectively "kick" the network out of a local energy trap.

With the help of this *guarded discrete stochastic* network (Adorf and Johnston, 1989; Johnston and Adorf, 1989a) we have been able to find solutions to the N-queens problem (Fig. 13.5) up to $N = 1024$ in modest time on a Texas Instruments Explorer workstation. This has to be compared to the results obtained by Stone and Stone (1987), who in their study of efficient search algorithms report that they have been unable to find solutions to the 97-queens problem. The network for the 1024-queens problem is very large: it consists of more than 10^6 neurons and more than 10^9 connections.

We have tested the GDS-net algorithm on a variety of other combinatorial problems, building on the fact that the network update rule is independent of the problem (all the problem specificities are encoded in the network connection strengths) and have started to analyze the reasons for its high performance. A number of contributing factors have been isolated, notably its built-in ability of "hypothetical evidential reasoning": good hypotheses are favoured and bad assumptions are likely to be retracted in the course of network evolution.

The exploitation of the potential of the GDS-net algorithm is the subject of on-going research. So far, it has found its most successful application in the area of automated telescope scheduling (see Johnston, 1989, and references therein).

13.2.2 Scheduling with ANNs

Having seen how the N-queens problem translates into a neural network makes it easy to understand how other combinatorial problems such as scheduling may be represented in a network structure. In abstract terms the scheduling problem consists of a set of activities to be arranged on a discretized timeline. In an immediate generalization of the square arrangement needed for the N-queens problem, the scheduling task can be represented in a rectangular "neural timetable": each activity is identified with a row and each time interval with a column of neurons (Fig. 13.6). A neuron being in its "on"-state shall mean that the activity corresponding to the neuron's row starts in the time interval corresponding to the neuron's column.

Absolute constraints such as "activity B may not start at time 3" can be implemented by applying a negative bias value to the corresponding neuron. Conversely, a scheduling *preference* can be introduced by a positive bias value. *Relative constraints* such as "activity A has to precede activity B" can easily be represented by inhibitory connections between the corresponding neuron rows (Fig. 13.6). *Resource constraints* can be taken into account by guard neurons

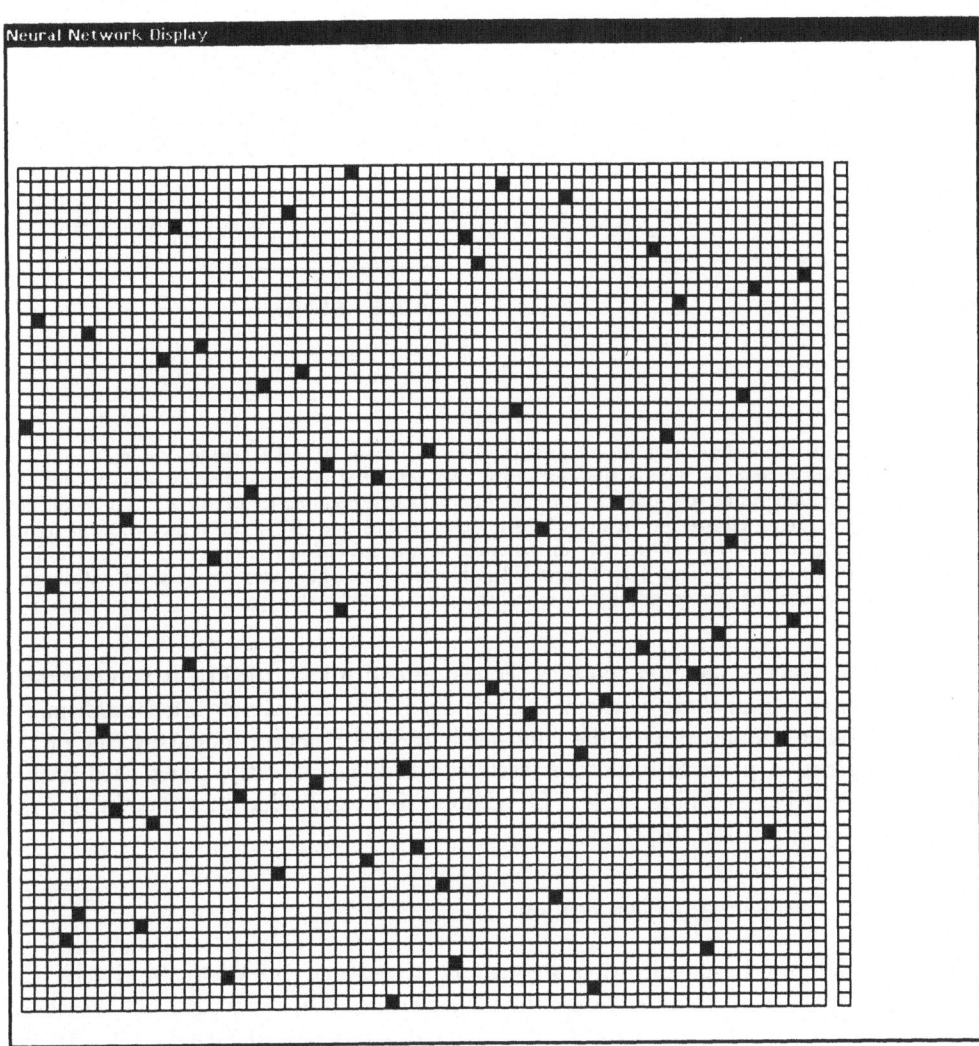

Fig. 13.5. Screen display of a solution to the 64-queens problem. The solution was found after 54 seconds by the "guarded discrete stochastic" (GDS) neural network algorithm implemented in Lisp on a Texas Instruments Explorer workstation.

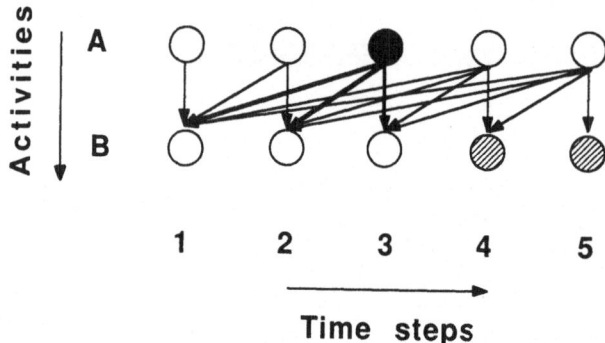

Fig. 13.6. Neural network for scheduling. Two activities, A and B, are shown which shall be placed on a timeline with 5 time intervals. An activated neuron indicates that the activity corresponding to its row starts in the time interval corresponding to its column. The hypothetical precedence constraint "A must be scheduled earlier than B" is taken into account by the inhibitory connections shown. If, for instance, neuron A3 (filled circle) is activated, it inhibits the activation of neurons B1, B2 and B3 (white circles). Inhibitory connections within each row (not shown) guarantee that each activity is scheduled once only.

which supervise neuron columns (Johnston and Adorf, 1989b). It is remarkable how natural all these constraints come out in the neural network formalism.

When combined with the controlled stochastic neuron updating rule of the GDS-net, a powerful scheduler results, which is currently being used within the SPIKE scheduling project (Johnston, 1989, and references therein) for the Hubble Space Telescope.

13.2.3 Unsupervised Classification

Our interest in classification algorithms stems from the desire to classify the objects in the IRAS Point Source Catalog which contains some $250,000$ records of observations with the space-bourne Infrared Astronomical Satellite (IRAS). We have considered subsets of these data (Adorf and Meurs, 1988; Meurs et al., 1988; Murtagh, 1988) which are characterized by a large number (10,000 to 30,000) of objects with a moderate number of features (4 to 6). Many of the catalogued objects are of unknown type, and within our unsupervised classification work we try to let the data "organize themselves" and see whether identifiable object categories emerge.

If we require the classes to be a disjoint decomposition of the set of objects, then the classification problem can be formally stated as follows: let $I = \{1, \ldots, n\}$ denote the set of objects, $\Omega = \{1, \ldots, o\}$ the set of classes and let i and ω label different objects and classes, respectively. We seek a clustering function $g : I \to \Omega$, such that $\omega = g(i)$ gives the class to which object i belongs.

Again, a rectangular neural network can be used to represent the classification task: identify each object with a neuron row and each class with a neuron column. A neuron being "on" indicates that the object corresponding to its row is a member of the class corresponding to its column (Fig. 13.7).

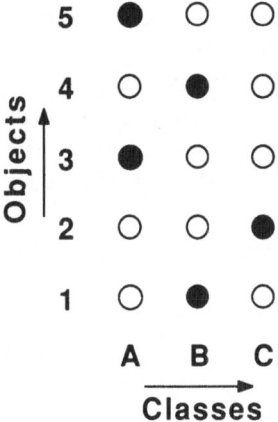

Fig. 13.7. Neural network for classification. Objects are identified with neuron rows, classes with neuron columns. The interconnects between the neurons (not shown) reflect the characteristics of the objective function to be used for the classification. The active neurons (filled circles) represent the hypothetical classification result: objects 1, 2, 3, 4 and 5 belong to classes A, B, A, C and B, respectively.

Among the many cluster algorithms which may be considered for unsupervised classification there are some which can be formulated as a combinatorial optimization problem with a symmetric bilinear objective function. If this function is interpreted as the Lyapunov ("energy") function of a Hopfield neural net, a network connection matrix can easily be read off and a classifier constructed (Adorf and Murtagh, 1988; 1989).

13.2.4 Perspective

13.2.4.1 What are the Near Term Prospects for Neurocomputing?. In the United States, where currently most research is done, the future of neurocomputing will — the usual sad story — largely depend on funding from military institutions such as DARPA. The provision of US $ 400 million to be spent over a couple of years on neural network research has been suggested — a decision is pending. Europe and Japan have initiated their own research programs, but the situation is less transparent.

A fairly safe prediction seems to be that artificial neural networks will most likely succeed in areas in which their biological counterparts work well, such as reasoning under uncertainty, processing of noisy data and control. The classical computer domains, such as numerical and symbolic computation, searching and sorting etc., where current serial processors easily outperform human brains, will continue to be served well by the von Neumann architecture.

It can be expected that the future will see more and more *trainable* (as opposed to *programmable*) systems, for which neural network models are prime candidates. Trainable systems are needed when it is unknown *how* a given task can be accomplished and when therefore no algorithm can be devised. However, in some areas neural networks have to compete with well-established and theoretically well-founded techniques, such as cross-correlation (see e.g. Adorf, 1986) or Bayesian approaches (see e.g. Duda and Hart, 1973) to supervised classification; work on performance comparisons is just starting.

It is most likely that neural networks will come to the non-specialist in the form of add-on boards to be plugged into conventional workstations or personal computers. Attention should be paid to developments in the optical computer area. Computers with optical rather then metallic communication pathways could provide the high connectivity which largely determines the power of neural networks.

13.2.4.2 Neural Networks in Astronomy. Apart from the applications of neural networks to scheduling and unsupervised classification described above, a neural network based approach is currently being investigated for *event location* in the COMPTEL γ-ray experiment (Simpson, 1987; Simpson and Reinhard, 1988) to be flown on NASA's Gamma Ray Observatory. First results show a substantial performance advantage over the previously used method, despite the fact that the neural network is only simulated on a serial processor.

Some other areas have been identified, in which neural networks might potentially be applied to astronomical problems (Adorf and Johnston, 1988). The most obvious of these are supervised *classification of spectra* or, more ambitiously, *morphological classification of galaxies*. A Hopfield net could be trained on classified examples and subsequently used as an associative, content addressable memory.

Real time mirror shape control will become very important for the adaptive optics (Merkle, 1988) of ESO's Very Large Telescope (see Ulrich, 1988) currently under construction. Controlling large active optics is a demanding task because of the large number of wavefront sensors and the frequency with which the actuator positions have to be updated. Here neural networks might offer simplicity combined with speed, in particular when implemented in an analogue rather than a digital fashion. An interesting question is whether a neural network could be trained to predict the atmospheric refractive behaviour in order to facilitate the control task.

Finally trainable neural networks may be beneficial for *querying* those *very large databases* to be assembled from the results of future astronomical space satellites such as Hubble Space Telescope, ROSAT, ISO and others. Actually, a system allowing "queries by example" would be nothing more than an immediate application of the previously mentioned supervised classification techniques. Imagine an "artificial retina" which, being shown a few examples of the kind of galaxies you are interested in, digs in the archive and finds all similar objects for you ...

13.3 Part III: Further Reading

Until very recently, research articles on neural networks were scattered among many journals, some of which are as widespread as *Nature* and *Science*, whereas others such as *Biological Cybernetics*, *Cognitive Science*, *Complex Systems* and the *Proceedings of the National Academy of Science* are not easily accessible at many places now interested in neurocomputing. References to neural network research articles, which due to the diversity of publications are not easy to find otherwise, can be searched with the help of *NeuralBase*, a 2700 entry annotated computer-readable bibliography, available from Anza Research, Cupertino. The bibliography can be queried either with a predefined keyword-based search algorithm or in conjunction with dBase III on PCs.

13.3.1 Magazines, Newsletters and Journals

The monthly *AI Expert* magazine, published by Miller Freeman Publications, and the quarterly *IEEE Expert*, published by the Computer Society of the IEEE, contain easy readers on artificial intelligence related developments (including neural networks) and numerous advertisements on relevant soft- and hardware products. These magazines can be recommended to those who have a cursory interest in connectionism and wish to keep updated.

Two neural network newsletters aim for quick turn-around times: *Intelligence: The Future of Computing*, founded by Edward Rosenberg, intends to be a place "to find all the news about neural networks". *Neurocomputers*, edited by Derek Stubbs and published bi-monthly by Gallifrey Publishing, is targeted at international communication of reports, notes, reviews, preprints and programs from the area of neural networks, neuroengineering and sixth generation computers.

Two journals focus on professional research in the area of neural systems: since early 1988 Pergamon Press issues *Neural Networks*, the official quarterly journal of the International Neural Network Society (for society information contact Dr. Harold Szu, Attn.: INNS, NRL-Code 5765, Washington, D.C. 20375-5000). The journal is co-edited by Shun-ichi Amari (Asia and Australia), Stephen Grossberg (North and South America) and Teuvo Kohonen (Europe and Africa) and publishes original contributions on brain and behaviour modelling and its applications to computing.

Terrence Sejnowski will act as editor-in-chief for *Neural Computing*, a quarterly journal first issued by MIT Press in Spring 1989. The journal will focus on short communications and research reviews from neuroscience, computer science, artificial intelligence, mathematics, physics, psychology, linguistics, adaptive systems, vision, speech, robotics, optical computing and VLSI.

Research articles on neural networks can also be found in more traditional AI journals such as in *Artificial Intelligence*, published by North-Holland, and in *Computational Intelligence*, published by the National Research Council Canada, and in journals such as *Computer, Vision, Graphics, and Image Processing*, *Pattern Recognition* and *Pattern Recognition Letters*.

Recently, the state-of-the-art in artificial neural network research has been reviewed in special issues of *Applied Optics* (**26**, No. 23, 1987) and *Computer* (**21**, No. 3, 1988).

13.3.2 Conference Proceedings

Important annual conferences on neuroscience are, in the USA:
- the *Annual Conference of the Cognitive Science Society,*
- the *Annual Meeting of the Society for Neuroscience,*
- the *IEEE Conference on Neural Information Processing Systems Natural and Synthetic,*
- the *IEEE International Conference on Neural Networks (ICNN)* and
- the *International Neural Network Society (INNS) Annual Meeting.*

The latter two conferences will be merged into the *International Joint Conference on Neural Networks* from 1989 onwards.

In Europe there are:
- the *Annual Meeting of the European Neuroscience Association (ENA),*
- the *European Conference on Neural Networks (nEuro)* and
- the *Neural Networks and Their Applications (Neuro-Nimes)* conference.

Carnegie-Mellon University provides a *Connectionist Models Summer School.* An up-to-date schedule of conferences on neural networks and related topics is maintained in the back of *Neural Networks.*

13.3.3 Books

Until very recently, only a few scholarly treatments of neural networks existed, most notably Marvin Minsky and Seymour Papert's *Perceptrons,* Michael Arbib's *Brains, Machines, and Mathematics,* and Teuvo Kohonen's *Self-Organization and Associative Memory.* These three books have recently been revised and reissued. *Perceptrons,* a classic work on threshold unit networks, was the first systematic study of parallel computational models; Minsky and Papert (1988) have added a chapter reviewing recent developments and identifying new research directions in neural networks. The "sparkling primer" *Brains, Machines, and Mathematics* (Arbib, 1987) is a very thoroughly written book from an expert who has stayed in the field for a long time. *Self-Organization and Associative Memory* (Kohonen, 1987b) is somewhat more difficult to read, but a must for someone who wants to use neural networks for unsupervised learning.

The two volume *Parallel Distributed Processing,* edited by David Rumelhart, James McClelland and the PDP Research Group, has been the first coherent treatment published in the recent wave of renewed interest in neural networks. *Parallel Distributed Processing* (Rumelhart et al., 1987; McClelland et al., 1987) is written as a series of fairly independent chapters, covering various aspects of computational neuroscience, with a slant towards psychological and biological models. Recently these volumes have been complemented by the workbook *Explorations in Parallel Distributed Processing* (McClelland and Rumelhart, 1988). James Anderson and Edward Rosenfeld (1988) have undertaken the invaluable effort of assembling a reader *Neurocomputing: Foundations of Research* with reprints of more than 40 important original research articles scattered among diverse journals. This book is highly recommended.

John Denker is the editor of the proceedings of the 1986 AIP Snowbird conference on *Neural Networks for Computing*. It contains more than 60 research articles covering the whole spectrum of neurocomputing issues.

Leo van Hemmen and Ingo Morgenstern have edited the proceedings of the 1986 *Heidelberg Colloquium on Glassy Dynamics*, which brought together physicists interested in spin glasses with researchers interested in combinatorial optimization and neural networks. An outstanding paper is that of Grötschel and collaborators on the computational complexity of calculating the ground state of a spin glass in the Ising model.

13.3.4 Introductions and Reviews

A very enjoyable short article, which may well serve as a first introduction to the field of neural computing, has been composed by Materna (1987), who then was with Hecht-Nielsen Neurocomputer Corporation, San Diego. He describes the astonishing "Kolmogorov's Neural Network Existence Theorem" and Grossberg's coupled differential network equations. Cited network applications include probability density function (PDF) estimation and the nearest neighbor classifier for pattern recognition. Unfortunately the article does not contain references.

Tank and Hopfield's (1987) article in *Scientific American* is also recommendable as an entry point to the area of "collective computations" based on neural networks.

Another very readable introduction to connectionism is the article of Fahlman and Hinton (1987). They describe in some depth the use of iterative (feedback) networks for constraint-satisfaction problems.

Caudill (1987, 1988a, b, c) has assembled a series of easy readers introducing the world of neurocomputing.

Lippmann (1987) provides a more extensive introduction to neural network computing. He compares six nets, among these the Hamming and the Hopfield net. The article is complemented by an extensive list of references.

A very interesting and readable account of some important events which influenced the historical development and commercialization of neurocomputing is given by Rosenfeld (1987).

Anderson (1986) briefly reviews connectionism and neural nets and compares the implementation effort of a conventional (DECtalk) and a neural net based (NETtalk) speech synthesizer.

Bridle (1987) discusses the application of neural networks to pattern recognition tasks. He relates "hidden Markov models" — the most successful approach in current speech recognition systems — with adaptive networks such as Rumelhart networks and Boltzmann machines, which "learn" from examples, e.g., how to do pattern classification. Twenty references complement this very readable article.

13.3.5 Theory and special networks

The two-state discrete neural network of the Hopfield type with asynchronous random updating was introduced by Hopfield (1982). A modification of this is a neural network based on continuous-state (analogue, graded) neurons with non-linear response and synchronous updating, also introduced by Hopfield (1984). Hopfield and Tank (1986) discuss the Lyapunov ("energy" function) method for understanding neural network behaviour. Earlier results from applying a Hopfield net to analogue-to-digital conversion and to the travelling salesman problem are summarized.

A number of networks and their potential application areas are briefly described in the *Advanced Neurocomputer Application Course* from Hecht-Nielsen Neurocomputer Corporation, San Diego.

Feldman (1987) thoroughly describes the Lyapunov ("energy") function method, which is useful for analyzing the dynamical behaviour of symmetric neural networks, in particular its convergence to fixed points. The article contains a large number of references to the original research literature.

Learning of connection weights and retrieval from associative memory is extensively discussed in Fogelman Soulie et al. (1987).

13.3.6 Applications and Implementations

A very interesting (and readable) article on neural computing is that of Hopfield and Tank in *Science* (1986). The authors explain how neural network algorithms can be applied to combinatorial optimization problems. One of their examples is the travelling salesman problem: if n is the number of cities, n^2 neurons are needed to represent the problem. For $n = 30$ cities for which there are about 10^{30} possible tours, 10^7 of which are "good", they claim that the Hopfield net, starting from a random trial solution, converges quickly to one of these.

Bounds (1987) exposes and compares various new approaches to *global optimization*, such as simulated annealing, genetic algorithms and artificial neural network processing. The work of Hopfield and Tank on the travelling salesman problem (TSP) is quoted. Bounds description of solutions to the TSP has, however, been severely criticized by Johnson (1987).

Szu and Messner (1986) use neural networks for adaptive scale and rotation invariant image recognition.

Wechsler (1987) discusses the use of neural networks in invariant object recognition tasks.

Ballard et al. (1983) discuss the use of neural nets for artificial vision. Particularly interesting is their remark on representing image features by neurons each of which measures the probability of "a hypothesis about the existence of a feature in the image" being correct.

Poggio et al. (1985) discuss the possibility of using neural nets for implementing variational principles in order to solve regularization problems.

Forrest et al. (1987) compare the implementation (software emulation) of neural network models on different hardware architectures, i.e. the Distributed Array Processor from ICL and the Meiko Computing Surface.

13.3.7 Hardware

North (1987) briefly reports on the "IEEE Conference on Neural Information Processing Systems — Neural and Synthetic" (Denver, 1987). Among other things he describes the Alspector-chip from Bell Laboratories containing the Hinton-Sejnowski neural network (also called "Boltzmann machine"), a stochastic generalization of the Hopfield network.

Within the IEEE Proceedings special issue on digital signal processing, Gass et al. (1987) describe the Texas Instruments Odyssey digital signal processing (DSP) board, which fits into an TI Explorer chassis. The work being done at the Computer Science Center of Texas Instruments, Dallas, on implementing the discrete Hopfield network is briefly mentioned. The Odyssey board contains four processors of TI's TMS320 family of digital signal processors and is described in Lin et al. (1987).

Hwang (1987) gives a general overview of supercomputing based on parallel architectures. A section on optical computing and neurocomputers mentions two neural circuits, namely the AT&T Bell Laboratories chip, containing 256 artificial neurons with over 64,000 possible interconnects, and a chip from TRW. Also TRW's Mark III virtual neural network machine is briefly described.

Acknowledgements. This work was carried out within the "Artificial Intelligence Pilot Project" at the ST-ECF. I am indebted to my colleague Mark Johnston at the Space Telescope Science Institute, Baltimore, for very fruitful and rewarding collaboration. I wish to thank Jeanette Rubner and Helge Ritter of the Technische Universität München for reviewing the draft manuscript and the editors of this book for continuous encouragement.

References

1. Aarts, E.H.L. and van Laarhoven, P.J.M. (1987), "Simulated annealing: a pedestrian review of the theory and some applications", in *Pattern Recognition Theory and Applications*, P.A. Devijver and J. Kittler (eds.), 179–192.
2. Ackley, D.H., Hinton, G.E. and Sejnowski, T.J. (1985), "A learning algorithm for Boltzmann machines", *Cognitive Science*, 9, 147–169.
3. Adorf, H.-M. (1986), "Classification of low-resolution stellar spectra via template matching — a simulation study", in *Data Analysis in Astronomy II*, V. Di Gesù, L. Scarsi, P. Crane, J.H. Friedman and S. Levialdi (eds.), Plenum Press, New York, 61–69.
4. Adorf, H.-M. and Johnston, M.D. (1987), "The Artificial Intelligence Pilot Project of the ST-ECF", *ST-ECF Newsletter*, 8, 4–5.
5. Adorf, H.-M. and Johnston, M.D. (1988), "Artificial neural nets in astronomy", in *Arbeitspapier der Gesellschaft für Mathematik und Datenverarbeitung*, 329, C. Lischka and J. Kindermann (eds.), GMD, Bonn, 3–5.
6. Adorf, H.-M. and Johnston, M.D. (1989), "A discrete stochastic 'neural network' algorithm for constraint satisfaction problems", in preparation.

7. Adorf, H.-M. and Meurs, E.J.A. (1988), "Supervised and unsupervised classification — the case of IRAS Point Sources", in *Large-Scale Structures in the Universe: Observational and Analytical Methods*, W.C. Seitter, H.W. Duerbeck and M. Tacke (eds.), Springer-Verlag, Heidelberg, 315–322.

8. Adorf, H.-M. and Murtagh, F. (1988), "Clustering based on neural network processing", in *COMPSTAT 88*, D. Edwards and N.E. Raun (eds.), Physica Verlag, Heidelberg, 239–244.

9. Adorf, H.-M. and Murtagh, F. (1989), "Unüberwachte Klassifikation mit einem diskreten, stochastischen 'neuronalen' Netzwerk", in preparation.

10. *Advanced Neurocomputer Application Course* (1987), Hecht-Nielsen Neurocomputer Corporation, San Diego.

11. Amit, D.J., Gutfreund, H. and Sompolinsky, H. (1985), *Physical Review A*, **32**, 1007.

12. Amit, D.J., Gutfreund, H. and Sompolinsky, H. (1985), "Storing infinite numbers of patterns in a spin-glass model of neural networks", *Physical Review Letters*, **55**, 1530–1533.

13. Anderson, J.A. (1986), "Networks for fun and profit", *Nature* **322**, 406–407.

14. Anderson, J.A. and Rosenfeld, E. (eds.) (1988), *Neurocomputing: Foundations of Research*, MIT Press/Bradford Books, Cambridge, MA.

15. Arbib, M.A. (1987), *Brains, Machines, and Mathematics*, Springer-Verlag Heidelberg.

16. Ballard, D.H., Hinton, G.E. and Sejnowski, T.J. (1983), "Parallel visual computation", *Nature*, **306**, 21–26.

17. Baum, E.B. (1986), "Towards practical neural computation for combinatorial optimization problems", in *Neural Networks for Computing*, J.S. Denker (ed.), 53–58.

18. Biema, van, M. (1987), "Parallelism in Lisp", *Proceedings of the 10th International Joint Conference on Artificial Intelligence*, Vol. 1, 56–61.

19. Blelloch, G. and Rosenberg, Ch.R. (1987), "Network learning on the Connection Machine", in *Proceedings of the 10th International Joint Conference on Artificial Intelligence*, Vol. 1, 323–326.

20. Bounds, D.G. (1987), "New optimization methods from physics and biology", *Nature*, **329**, 215–219.

21. Bridle, J.S. (1987), "Adaptive networks and speech pattern processing", in *Pattern Recognition Theory and Applications*, P.A. Devijver and J. Kittler (eds.), 211–222.

22. Campbell, P.J. (1977), "Gauss and the Eight Queens Problem: a study in miniature of the propagation of historical error", *Historia Mathematica*, **4**, 397–404.

23. Carpenter, G.A. and Grossberg, S. (1988), "The art of adaptive pattern recognition by a self-organizing neural network", *Computer*, **21**, 77–88.

24. Caudill, M. (1987), "Neural networks primer, part I", *AI Expert*, **2**, No. 12, 46–52.

25. Caudill, M. (1988a), "Neural networks primer, part II", *AI Expert*, **3**, No. 2, 55–61.

26. Caudill, M. (1988b), "Neural networks primer, part III", *AI Expert*, **3**, No. 6, 53–59.

27. Caudill, M. (1988c), "Neural networks primer, part IV", *AI Expert*, **3**, No. 8, 61–67.

28. Caudill, M. and Butler, Ch. (eds.) (1987), *IEEE First International Conference on Neural Networks*, Institute of Electrical and Electronic Engineers, Service Center, Piscataway, NJ.

29. Cook, S.A. (1983), "An overview of computational complexity", *Communications of the ACM*, **26**, 401–408.

30. Davis, L. (ed.) (1987), *Genetic Algorithms and Simulated Annealing*, Pitman, London.

31. Denker, J.S. (ed.) (1987), *Neural Networks for Computing*, AIP Conference Proceedings No. 151, Snowbird, UT, American Institute of Physics, New York.

32. Devijver, P.A. and Kittler, J. (eds.) (1987), *Pattern Recognition Theory and Applications*, NATO ASI F30, Springer-Verlag, Heidelberg.

33. Diederich, J. (1988), "Trends im Konnektionismus", *KI* **1**, 28–32.

34. Diederich, J. and Lischka, Ch. (1987), "SPREAD-3, Ein Werkzeug zur Simulation konnektionistischer Modelle auf Lisp-Maschinen", *Rundbrief des FA 1.2 Künstliche Intelligenz und Mustererkennung in der Gesellschaft für Informatik*, **45**, 75–82.

35. Duda, R.O. and Hart, P.E. (1973), *Pattern Classification and Scene Analysis*, Wiley, New York.

36. Eklundh, J.O., Lansner, A. and Wessblad, R. (1986), "Classification of multispectral images using associative nets", in *Proceedings of the 8th International Conference on Pattern Recognition*, IEEE Computer Society Press, New York, 1240–1243.

37. Fahlman, S.E. and Hinton, G.E. (1987), "Connectionist architectures for artificial intelligence", *Computer*, **20**, 100–109.

38. Fanty, M.A. (1986), "Context-free parsing with connectionist models", in *Neural Networks for Computing*, J.S. Denker (ed.), 140–145.

39. Farhat, N.H. (1987), "Optoelectronic analogs of self-programming neural nets: architecture and methodologies for implementing fast stochastic learning by simulated annealing", *Applied Optics*, **26**, 5093–5103.

40. Feldman, J.A. (1985), "Energy and the behavior of connectionist models", Computer Science Department, University of Rochester, Technical Report 155, 1–39.

41. Feldman, J.A. (1987), "Energy methods in connectionist modelling", in *Pattern Recognition Theory and Applications*, P.A. Devijver and J. Kittler (eds.), 223–247.

42. Fogelman Soulie, F., Gallinari, P. and Thiria, S. (1987), "Learning and associative memory", in *Pattern Recognition Theory and Applications*, P.A. Devijver and J. Kittler (eds.), 249–268.

43. Foo, Y.-P.S. and Takefuji, Y. (1988a), "Stochastic neural networks for solving job-shop scheduling: Part 1. Problem representation", in *Proceedings of the IEEE Second International Conference on Neural Networks (ICNN)*, Vol. II, 275–282.

44. Foo, Y.-P.S. and Takefuji, Y. (1988b), "Stochastic neural networks for solving job-shop scheduling: Part 2. Architecture and simulations", in *Proceedings*

of the IEEE Second International Conference on Neural Networks (ICNN), Vol. II, 283–290.

45. Forrest, B.M., Roweth, D., Stroud, N., Wallace, D.J. and Wilson, G.V. (1987), "Implementing neural network models on parallel computers", *Computer Journal*, **30**, 413–419.

46. Fukushima, K. (1988a), "A neural network for visual pattern recognition", *Computer*, **21**, No. 3, 65–75.

47. Fukushima, K. (1988b), "Neocognitron: a hierarchical neural network capable of visual pattern recognition", *Neural Networks*, **1**, 119–130.

48. Gallant, S.I. (1988), "Connectionist expert systems", *Communications of the ACM*, **31**, 152–169.

49. Garey, M.R. and Johnson, D.S. (1979), *Computers and Intractability*, W.H. Freeman and Company, San Francisco.

50. Gass, W.S., Tarrant, R.T., Pawate, B.I., Gammel, M., Rajasekaran, P.K., et al. (1987), "Multiple digital signal processor environment for intelligent signal processing", *Proceedings of the IEEE*, **75**, 1246.

51. Geman, S. and Geman, D. (1984), "Stochastic relaxation, Gibbs distributions, and the Bayesian restoration of images", *IEEE Transactions on Pattern Analysis and Machine Intelligence*, **PAMI-6**, 721–741.

52. Genthner, A., Horner, H., Lange, R. and Männer, R., "NERV — ein Simulationssystem für neuronale Netzwerke: Software und Benutzerinterface", *Sprachen, Algorithmen und Architekturen für Parallelrechner*, Gesellschaft für Informatik, Fachgruppen 2.1.4 und 3.1.2, 315–329.

53. Goddard, N., Lynne, K. and Mintz, T. (1987), "The Rochester Connectionist Simulator", Technical Report 233, Computer Science Department, University of Rochester. Software distributed by: Ms. Rose Peet, Computer Science Department, University of Rochester, New York.

54. Goles, E. and Vichiniac, G.Y. (1986), "Lyapunov functions for parallel neural networks", in *Neural Networks for Computing*, J.S. Denker (ed.), 165–181.

55. Graf, H.P., Jackel, L.D., Howard, R.E., Straughn, B., Denker, J.S., Hubbard, W., Tennant, D.M. and Schwartz, D. (1986), "VLSI implementation of a neural network memory with several hundreds of neurons", in *Neural Networks for Computing*, J.S. Denker (ed.), 182–187.

56. Graf, H.P., Jackel, L.D. and Hubbard, W.E. (1988), "VLSI implementation of a neural network model", *Computer*, **21**, No. 3, 41–49.

57. Grötschel, M., Jünger, M. and Reinelt, G. (1986), "Calculating exact ground states of spin glasses: a polyhedral approach", in *Heidelberg Colloquium on Glassy Dynamics*, J.L. van Hemmen and I. Morgenstern (eds.), 325–353.

58. Hecht-Nielsen, R. (1987), "Komolgorov's mapping neural network existence theorem", in *Proceedings of the IEEE First International Conference on Neural Networks (ICNN)*, Vol. III, 11–14.

59. Hemmen, van, J.L. and Morgenstern, I. (1987), *Heidelberg Colloquium on Glassy Dynamics*, Springer-Verlag, Heidelberg.

60. Hernandéz, D. (1988), "Bericht über den Workshop 'Konnektionismus' bei der GMD", *KI 3/88*, 12–13.

61. Hertz, J.A., Grinstein, G. and Solla, S.A. (1986), "Irreversible spin glasses and neural networks", in *Heidelberg Colloquium on Glassy Dynamics*, J.L. van Hemmen and I. Morgestern (eds), 538–546.

62. Hillis, W.D. and Barnes, J. (1987), "Programming a highly parallel computer", *Nature*, **326**, 27–30.

63. Hinton, G.E. (1987), "Connectionist learning procedures", Carnegie-Mellon University, Department of Computer Science, Technical Report CMU-CS-87-115, 1–32.

64. Hinton, G.E. and Sejnowski, T.J. (1987), "Learning and relearning in Boltzmann machines", in *Parallel Distributed Processing*, D.E. Rumelhart, J.L. McClelland and the PDP Research Group (eds.), Vol. 1, 282–317.

65. Hirsch, M.W. (1987), "Convergence in neural nets", in *Proceedings of the IEEE First International Conference on Neural Networks (ICNN)*, M. Caudill and C. Butler (eds.), Vol. II, 115–125.

66. HNC (1988), "System Specifications", Hecht-Nielsen Neurocomputer Corporation, 5893 Oberlin Drive, San Diego, CA 92121, USA.

67. Hoffmann,E.J., Loessi, J.C. and Moore, R.C. (1969), "Constructions for the solution of the m Queens Problem", *Mathematics Magazine*, **42**, 66–72.

68. Hopfield, J.J. (1982), "Neural networks and physical systems with emergent collective computational abilities", *Proceedings of the National Academy of Sciences*, **79**, 2554–2558.

69. Hopfield, J.J. (1984), "Neurons with graded response have collective computational properties like those of two-state neurons", *Proceedings of the National Academy of Sciences*, **81**, 3088–3092.

70. Hopfield, J.J. and Tank, D.W. (1985), ""Neural" computation of decisions in optimization problems", *Biological Cybernetics*, **52**, 141–152.

71. Hopfield, J.J. and Tank, D.W. (1986), "Computing with neural circuits: a model", *Science*, **233**, 625–633.

72. Hutchinson, J., Koch, Ch., Luo, J. and Mead, C. (1988), "Computing motion using analog and binary resistive networks", *Computer*, **21**, No. 3, 52–63.

73. Hwang, K. (1987), "Advanced parallel processing with supercomputer architectures", *Proceedings of the IEEE*, **75**, 1348–1379.

74. Hwang, K., Ghosh, J. and Chowkwanyun, R. (1987), "Computer architectures for artificial intelligence processing", *Computer*, **20**, No. 1, 19–27.

75. Jeffrey, W. and Rosner, R. (1986), "Optimization algorithms: simulated annealing and neural network processing", *Astrophysical Journal*, **310**, 473–481.

76. Johnson, D. (1987), "More approaches to the travelling salesman guide", *Nature*, **330**, 525–525.

77. Johnston, M.D. (1989), "Knowledge based telescope scheduling", this volume.

78. Johnston, M.D. and Adorf, H.-M. (1989a), "Stochastic neural networks for constraint satisfaction problems", in preparation.

79. Johnston, M.D. and Adorf, H.-M. (1989b), "Scheduling with neural networks", in preparation.

80. Karp, R.M. (1986), "Combinatorics, complexity, and randomness", *Communications of the ACM*, **29**, 98–109.

81. Karp, R.M. and Widgerson, A. (1985), "A fast parallel algorithm for the maximal independent set problem", *Journal of the ACM*, **32**, 762–773.

82. Kibler, D.F. and Conery, J. (1985), "Parallelism in AI programs", *Proceedings of the 9th International Joint Conference on Artificial Intelligence*, Vol. 1, 53–56.

83. Kirkpatrick, S., Gelatt, C. and Vecchi, M. (1983), "Optimization by simulated annealing", *Science* **22**, 671–680.

84. Kohonen, T. (1987a), "Adaptive, associative, and self-organizing functions in neural computing", *Applied Optics*, **26**, No. 23, 4910–4918.

85. Kohonen, T. (1987b), *Self-Organization and Associative Memory*, Second Edition, Springer-Verlag, Heidelberg.

86. Kohonen, T. (1988), "The 'neural' phonetic typewriter", *Computer*, **21**, 11–22.

87. Kolmogorov, A.N. (1957), "On the representation of continuous functions of many variables by superposition", *Dokl. Akad. Nauk USSR*, **114**, 953–956.

88. Kumar, V. and Kanal, L.N. (1984), "Parallel branch-and-bound formulations for AND/OR tree search", *IEEE Transactions on Pattern Analysis and Machine Intelligence*, **PAMI-6**, 768–778.

89. Lapedes, A. and Farber, R. (1986), "Programming a massively parallel, computation universal system: static behavior", in *Neural Networks for Computing*, J.S. Denker (ed.), 283–298.

90. Le Cun, Y. (1985), "A learning scheme for asymmetric threshold networks", in *Proc. Cognitiva*, **85**, Paris, 599–604.

91. Linsker, R. (1988), "Self-organization in a perceptual network", *Computer*, **21**, No. 3, 105–117.

92. Lin, K.-S., Frants, G.A. and Simar, R. (1987), "The TMS320 family of Digital Signal Processors", *Proceedings of the IEEE*, **75**, 1143–1159.

93. Lippmann, R.P. (1987), "An introduction to computing with neural nets", *IEEE Acoustics, Speech, and Signal Processing Magazine*, 4–22.

94. Materna, T. (1987), "Neural networks enter high speed marketplace", *Computer Technology Review*, **VII**, No. 7.

95. McClelland, J.L. and Rumelhart, D.E. (1988), *Explorations in Parallel Distributed Processing, A Handbook of Models, Programs, and Exercises*, MIT Press/Bradford Books, Cambridge, MA.

96. McClelland, J.L. Rumelhart, D.E., and the PDP Research Group (1987), *Parallel Distributed Processing, Explorations in the Microstructure of Cognition, Vol. 2: Psychological and Biological Models*, MIT Press/Bradford Books, Cambridge, MA.

97. McCulloch, W.S. and Pitts, W.H. (1943), "A logical calculus of the ideas immanent in nervous activity", *Bull. Math. Biophys.*, **5**, 115–133.

98. Merkle, F. (1988), "The VLT adaptive optics prototype system: status May 1988", *The Messenger*, **52**, 5–7.

99. Meurs, E.J.A., Adorf, H.-M. and Harmon, R.T. (1988), "Mapping the extragalactic sky with IRAS PSC data", in *Astronomy from Large Databases: Scientific Objectives and Methodological Approaches*, F. Murtagh and A. Heck (eds.), European Southern Observatory, Garching bei München, 49–54.

100. Mézard, M. (1986), "Spin glasses and optimization", in *Heidelberg Colloquium on Glassy Dynamics*, J.L. van Hemmen and I. Morgenstern (eds.), Springer-Verlag, Heidelberg, 354–372.

101. Minsky, M.L. and Papert, S. (1988), *Perceptrons* (Expanded Edition), MIT Press/Bradford Books, Cambridge, MA.

102. Mueller, P. and Lazzaro, J. (1986), "Real time speech recognition", in *Neural Networks for Computing*, J.S. Denker (ed.), 321–326.

103. Murtagh, F. (1988), "Multivariate data analysis: background and example", in *Large-Scale Structures in the Universe: Observational and Analytical Methods*, W.C. Seitter, H.W. Duerbeck and M. Tacke (eds.), Springer-Verlag, Heidelberg, 308–314.

104. North, G. (1987), "Implementation and analysis", *Nature*, **330**, 522–523.

105. Paik, E., Gunger, D. and Skrzypek, J. (1987), "UCLA SFINX — a neural network simulation environment", in *Proceedings of the IEEE First International Conference on Neural Networks (ICNN)*, Vol. III, 367–375.

106. Parker, D.B. (1985), "Learning-logic", Technical Report TR-47, Sloan School of Management, MIT, Cambridge, MA.

107. Platt, J.C. and Hopfield, J.J. (1986), "Analog decoding using neural networks", in *Neural Networks for Computing*, J.S. Denker (ed.), 364–369.

108. Poggio, T., Torre, V. and Koch, Ch. (1985), "Computational vision and regularization theory", *Nature*, **317**, 314–319.

109. Ramanujam, J. and Sadayappan, P. (1988), "Optimization by neural networks", in *Proceedings of the IEEE Second International Conference on Neural Networks (ICNN)*, Vol. II, 325–332.

110. Ritter, H., Martinetz, T.M. and Schulten, K. (1989), "Topology-conserving maps for learning visuomotor-coordination", *Neural Networks*, in press.

111. Ritter, H. and Schulten, K. (1987), "Kohonen's self-organization maps: exploring their computational capabilities", in *Proceedings of the IEEE First International Conference on Neural Networks (ICNN)*, 109–116.

112. Rosati, J., Shepanski, J. and Skrzypek, J. (1987), "Artificial neural systems", State of the Art Limited, Documentation, 1–207.

113. Rosenfeld, E. (1987), "Neurocomputing — a new industry", in *Proceedings of the IEEE First International Conference on Neural Networks (ICNN)*, Vol. IV, 831–838.

114. Rumelhart, D.E., McClelland, J.L., and the PDP Research Group (1987), *Parallel Distributed Processing, Explorations in the Microstructure of Cognition, Vol. 1: Foundations*, MIT Press, Cambridge, Massachusetts.

115. SAIC (1988), "System Specifications", Sigma Neurocomputer Systems Division, Scientific Applications International Corporation, 10260 Campus Point Drive, MS 34, San Diego, CA 92121, USA.

116. Schwartz, T.J. (1988), "12-Product wrap-up: neural networks", *AI Expert*, **3**, No. 8, 73–85.

117. Sejnowski, T.J. (1987), "Parallel networks learn to pronounce English text", *Complex Systems*, **1**, 145–168.

118. Sejnowski, T.J. and Rosenberg, Ch.R. (1986), "NETtalk: a parallel network that learns to read aloud", Technical Report JHU/EECS-86/01, Johns Hopkins University, Baltimore, MD.

119. Shapiro, S.F. (1988), "Robotic systems learn through experience", *Computer Design*, **27**, No. 20, 54–68.

120. Simpson, G. (1987), "Event-location using neural nets", COMPTEL Experiment, Technical Report COM-RP-UNH-F70-038, 1–14.

121. Simpson, G. and Reinhard, K. (1988), "A new approach to event location", COMPTEL Experiment, Technical Report COM-TN-UNH-F70-044, 1–18.

122. Sivilotti, M.A., Emerling, M.R. and Mead, C.A. (1986), "VLSI architecture for implementation of neural networks", in *Neural Networks for Computing*, J.S. Denker (ed.), 408–413.

123. Stone, H.A. and Stone, J.M. (1987), "Efficient search techniques — an empirical study of the N-queens problem", *IBM Journal of Research and Development*, **31**, 464–474.

124. Szu, H. (1986), "Fast simulated annealing", in *Neural Networks for Computing*, J.S. Denker (ed.), 420–425.

125. Szu, H.H. and Messner, R.A. (1986), "Adaptive invariant novelty filters", *Proceedings of the IEEE*, **74**, No. 3, 518–519.

126. Tagliarini, G.A. (1987), "Solving constraint satisfaction problems with neural networks", in *Proceedings of the IEEE First International Conference on Neural Networks (ICNN)*, Vol. III, 741–747.

127. Tanimoto, S. (1986), "Trends in parallel processing applications (panel discussion)", in *Data Analysis in Astronomy II*, V. Di Gesù, L. Scarsi, P. Crane, J.H. Friedman and S. Levialdi (eds.), Plenum Press, New York, 377–395.

128. Tank, D.W. and Hopfield, J.J. (1986), "Simple 'neural' optimization networks: an A/D converter, signal decision circuit, and a linear programming circuit", *IEEE Transactions on Circuits and Systems*, **CAS-33**, 533–541.

129. Tank, D.W. and Hopfield, J.J. (1987), "Collective computation in neuronlike circuits", *Scientific American*, **257**, No. 6, 62–70.

130. Thakoor, A.P., Moopenn, A., Lambe, J. and Khanna, S.K. (1987), "Electronic hardware implementation of neural networks", *Applied Optics*, **26**, No. 23, 5085–5092.

131. Touretzky, D.S. and Hinton, G.E. (1986), "A distributed connectionist production system", Technical Report CMU-CS-86-172, 38 pp, Carnegie-Mellon University, Department of Computer Science.

132. Ulrich, M.-H. (ed.) (1988), *ESO Conference on Very Large Telescopes and their Instrumentation*, 2 vols., European Southern Observatory, Garching bei München.

133. Waltz, D.L. (1987), "Applications of the Connection Machine", *Computer*, **20**, No. 1, 85–97.

134. Watrous, R.L. and Shastri, L. (1987), "Learning phonetic features using connectionist networks", in *Proceedings of the 10th International Joint Conference on Artificial Intelligence*, Vol. II, 851–857.

135. Wechsler, H. (1987), "Network representations and match filters for invariant object recognition", in *Pattern Recognition Theory and Applications*, P.A. Devijver and J. Kittler (eds.), 269–276.

136. Werbos, P.J. (1974), "Beyond regression: new tools for prediction and analysis in the behavioral sciences", PhD thesis, Harvard University, Cambridge, MA.

137. Widrow, B. and Winter, R. (1988), "Neural nets for adaptive filtering and adaptive pattern recognition", *Computer*, **21**, No. 3, 25–39.

138. Willems, J.L. (1970), *Stability Theory of Dynamical Systems*, Thomas Nelson, London.

139. Zipser, D. and Rabin, D.E. (1987), "P3: a parallel network simulation system", in *Parallel Distributed Processing*, D.E. Rumelhart, J.L. McClelland, and the PDP Research Group (eds.), Vol. 1, 488–506.

14 Applications of AI in Astronomy: A View Towards the Future

R. Albrecht*
Space Telescope — European Coordinating Facility
European Southern Observatory
Karl-Schwarzschild-Str. 2
D-8046 Garching bei München
F.R. Germany

14.1 Introduction

Much has been written about the enormous growth of our computing capabilities during the last three decades; and it has been astounding indeed. However, most of this growth has to be attributed to improvements of hardware technology (Albrecht et al., 1986). Software techniques, on the other hand, have only recently begun to change, for two reasons: primarily because the conventional software development paradigm has become a bottleneck of software production, and secondly because new software techniques are required to take full advantage of advanced hardware solutions. It is interesting to note that some of these advanced software concepts were conceived immediately after computers became possible to build — the field of Artificial Intelligence (AI), for instance, has been a bona fide subfield of computer science since 1956.

Computers were first used in astronomy relatively late; today they are as much part of astronomy as the telescope is. However, most of what astronomers do with computers today is not fundamentally different from what was done in the late sixties. Thus, after the first innovation through the introduction of computers to the field the time has come to again examine the field of computer science for innovative techniques which might usefully be employed in astronomy.

---

*Affiliated to the Astrophysics Division, Space Science Department, European Space Agency.

One of the most promising new techniques is the area of Artificial Intelligence which emerged during the last five years.

AI will impact the science of astronomy in more than one way. Obviously, we will enjoy the advantages of improved computational tools, which will enable us to tackle complex tasks and to solve problems which could not previously be solved. There is, however, another aspect, which will become important on a somewhat longer timescale: AI will change the way we use computers and at the same time it will change the way we go about collecting and processing information. Finally, and most importantly, AI has the potential of changing the methodology which we employ in doing science, thus enabling us to gain radically new insights into the universe.

14.2 The Issue of Intelligence

This is the point where usually definitions of "intelligence" and justifications for "artificial" are being listed. Readers who have successfully progressed through this book to this paper will have seen such definitions in other contributions.

Let us instead try to look at the issue of intelligence in a more general manner. Humans are intelligent by definition; in fact most people will agree that mammals in general display intelligent behaviour, sometimes to a surprising degree. Insect intelligence is harder to understand, but undoubtedly present and indeed necessary — an anthill, for instance, is too complex a system to be operated without intelligence.

Taking another step we find that some of the behaviour displayed by plants is fairly intelligent: the way they turn their leaves towards the light, how they protect themselves from heat and cold, and so on. In plants we have been successful in discovering some of the chemical processes which govern this behaviour — but why should the fact that we found out how it works remove the intelligence from such actions?

In short, intelligence is not a property unique to the human brain, nor does its manifestation necessarily require human nervous processes. In view of this it should not be surprising that sophisticated electronic systems, properly programmed, should be able to display a quite considerable degree of what can be classified as intelligent behavior.

There is, however, a fundamental difference between machine intelligence and human intelligence: while the results of applying either kind to a problem might be the same, or at least both results might be acceptable, in addition to the intellectual effort the human will experience him/herself as an acting individual, able to contemplate the effects of the solution just discovered on the discoverer. Machine intelligence, on the other hand, does not need, nor does it seek, to simulate the elements "consciousness" and "self".

14.3 Current AI Developments

AI is multi-facetted and AI development is going on in so many places and in so many areas that it is quite impossible to present a comprehensive overview. A recent survey conducted by one of the leading trade magazines, for instance, found more than 100 expert systems in production use by Japanese companies. So what we want to do instead is to identify the main reasons for and the main thrust of ongoing developments in order to be able to assess their probable impacts on astronomy.

AI based systems can most usefully be employed in solving complex problems. Complex, as opposed to complicated, means that the different aspects of the problem are easy to understand, but that they interact in multiple ways. In other words, conditions change as functions of other conditions, which in turn are again contingent on other events. Experiments have shown that humans perform poorly trying to solve problems with more than two such layers of complexity.

Within the last two decades the planning and implementation of technological projects and the organizations around them have grown in complexity to an extent that quite a few of these projects have now exceeded the abilities of individuals to maintain an overview and to base decisions on a comprehensive understanding of the boundary conditions and an awareness of all possible ramifications. This is, in particular, true for large space projects. Since these projects will not become less complex the only solution will be to make increased use of machine based reasoning and knowledge processing. For this reason we will see an increase of the use of knowledge based systems in industry in areas like project management, planning and scheduling, quality control, diagnostics and maintenance, etc. The result of this development will be that the technology will become available at modest prices and in a form which makes it usable by non-computer experts.

14.3.1 Expert Systems

The increased need to acquire and to process knowledge has lead to research into how to best represent, store and manipulate knowledge through computers. This has to be seen in combination with the enormous growth of computer communication facilities and data base technology, which makes it possible to access required information regardless of where it resides.

Within the last five years expert systems have changed from cumbersome toys which were only used on test problems in research laboratories to production tools used for everything from fault diagnosis of automobile engines to stock trading on Wall Street. Expert system shells are commercially available, their capabilities and operational speed continue to grow while the prices are going down.

While this development *per se* does not impact astronomy, it will, however, do so in the long run by making tools available which can be applied to problems relevant to astronomy. We have already seen similar developments in other areas, for example in the area of low signal level opto-electronic detectors, where tremendous progress in astronomy was made by utilizing technology originally developed for the military.

14.3.2 Artificial Neural Networks

Artificial Neural Nets is another candidate for promising new technology coming out of AI related developments. While the concept had been around for decades it was only recently that the potential was fully recognized.

Detailed descriptions of Artificial Neural Networks and their oparation can be found elsewhere (Hopfield and Tank, 1986; Lippmann, 1987; Adorf, 1989). Basically the way the net operates is that it turns input stimuli into output signals in a manner which is very similar to the way in which biological neural systems work; thus the name. A new discipline of Computer Science is emerging around the concept of Neural Nets: the discipline of Connectionism.

Possible applications of Neural Net technology are in areas where solutions have to be found in the presence of ambiguous and conflicting information, like in deciphering handwriting or understanding speech; or when a set of input stimuli allows more than one viable solution, like scheduling a spacecraft (Miller et. al., 1988). Promising applications in astronomy also include machine assisted classification (Adorf and Murtagh, 1988).

Artificial Neural Networks lend themselves to implementation on massively parallel hardware, which is not only computationally interesting, but will also lead to the development of panoramic detectors with Neural Networks attached to them, which will not only be able to produce an image, but will also be capable of instantly parametrizing (recognising) it.

Expert systems will also benefit from neural networks: they will be employed in rule base searching, conflict resolution, reasoning with uncertainty, and rule firing strategy (although the connectionists already call expert systems "good old fashioned AI").

14.4 Short Term Developments

Most AI development in astronomy right now is based on expert system technology. The most promising applications are the automatic classification of large bodies of data (Rampazzo et al., 1988). In keeping with the title of this paper we will not dwell on these.

However, expert systems can be used in other ways. The amount of software available to the astronomer has grown enormously, to the extent that some visitors at the European Southern Observatory are not complaining any more about the lack of available options, but about their multitude and complexity.

In fact, it is a definite indication of the need for improvement to realize that there are still people who take their plots to the draughtsperson and wait several days to get the results, instead of using one of the several plot packages which are available on the computer. Similarly, there are still people who use the printer-terminal in the library to get data from SIMBAD, most often then turning around and typing the information into a computer. They could have shortcircuited the process by using STARCAT and networking software, except that — right or wrong — they feel that immersing themselves into the "computerese" which is necessary to utilize those tools would take their attention away from the (astronomical) problem at hand.

For these reasons the desirability of "user friendliness" has been proclaimed. Indeed, many packages offer extensive on-line help as well as nice hard-copy documentation; command languages and user interfaces of various degrees of complexity and functionality abound. However, not only is the sheer volume of some of the manuals frightening, the documentation of the different systems follows different and inconsistent guidelines. In most cases, one has to pretty much know what one wants and how to do it in order to successfully use the system.

In addition to the problems mentioned above, there are significant areas of astronomical research in which we do not use the computer at all, although the technology required is already available and the capabilities could be implemented easily. One of the most obvious areas is library work, where so far we are using the computer only in order to generate lists of preprints and for some administrative work. Obvious simple improvements, already being offered through commercial packages, are alerting services based on titles and keywords.

14.4.1 The Concept of the Computer-Based Research Assistant

The only viable solution to the problem of how to use the computer more efficiently is to get the computer to help us.

Trying to define a mental model of the kind of help which would be ideal, one quickly finds that the most ideal help would be something resembling a Research Assistant. The ideal (human) research assistants are of senior undergraduate or junior graduate level, so they be talked to in domain terminology (jargon). They know, for instance, that Right Ascension and Magnitudes run "upside down", and why. They can be asked to go look something up in the library, and they can be relied on that they also consider the context correctly. We expect them to provide unsolicited input if relevant, and to check our reasoning. Most importantly, they should know how to get a publication quality plot out of the computer, how to run the data analysis system, or how to contact our colleagues in California through a computer network. Another quality which the research assistant has to possess is superb communication skill.

Not surprisingly, the "assistant" concept is being used in a number of areas with similar requirements: "mudman" is a software system for support and evaluation of geology tests, "pilots assistant" is a system which aids pilots in the operation of complex aircraft, "programmers apprentice" controls the functions of a software development environment.

We have done the first experiments here: the Data Analysis Assistant can perform the generic calibration of CCD frames and map the operation into one of two possible data analysis systems: either into MIDAS (Munich Image Data Analysis System, used at ESO/ ST-ECF), or into STSDAS (Space Telescope Science Data Analysis System, used at the STScI) (Johnston, 1987).

14.4.2 Possible Trial Implementation

The area in which the single most important gain for the largest number of people can be achieved is the area of data analysis support.

The data analysis system (for example the MIDAS system) could be covered with an expert system driven user interface, which has access to all MIDAS documentation and which shields the user from the complexities of the MIDAS command language. The system should have limited natural language understanding capabilities in order to be able to extract out of the user the information which MIDAS needs to successfully perform the required data analysis operations.

This can be achieved relatively simply by replacing the command input terminals of the MIDAS work stations with PC class machines. Using window-, icon-, mouse- or pointer- (WIMP) interface technology the user would be free to either talk directly to MIDAS (or to other tools, for that matter), or to use varying degrees of expert help through the "Assistant".

As we gain more experience with this approach, we can tie more and more of the tools which are currently residing on the computer into this system: query operations on large data bases, network access, etc. It goes without saying that alternative data analysis systems such as AIPS (Astronomical Image Processing System, developed at NRAO) or IRAF (Interactive Reduction and Analysis Facility, developed at NOAO) could be made available through such a user interface in a painless (for the user) manner.

The result of this would be the possibility for the casual user to optimally exploit all available tools: in the end, most users are casual users, in the sense that even an expert on MIDAS will need help trying to use AIPS, or STARCAT (Space Telescope Archive and Catalogue, the user interface to the ESO/ST-ECF Science Data Archive). An intelligent interface of the sort described above has additional important advantages: for instance, it would be possible to maintain MIDAS as is, i.e. it can be used on computers which do not have sophisticated user terminals; systems could cooperate user-invisibly underneath the interface layer without the need of making changes in these systems.

14.4.3 Optimum Calibration of HST Data

Observing time on HST will be an extremely scarce resource. For this reason it is foreseen to maximise the use of the HST Science Data Archive.

While HST data will be calibrated before being placed into the Archive we know from our experience with previous space based telescopes that the calibration which is being performed in near real time is not the best calibration, especially early in the lifetime of a mission. The consequence of this is that all data which get extracted out of the Archive at a later time will have to be recalibrated. This process, however, requires not only a thorough understanding of the HST and the Science Instruments, it also requires detailed knowledge of the calibration history of the different instruments and their multiple modes of operation, as well as the changes with time. The only viable solution to this problem is to capture the required knowledge in a knowledge base and use expert system technology to design a suitable generic calibration procedure and to subsequently map it into a target data analysis system.

14.4.4 The Scientific Library

Not much has changed in several centuries in our scientific libraries: they are basically a loosely organized repository of hard copy material. "Computerizing" a library has so far only meant machine assisted administrative procedures.

While reading incoming journals is relatively enjoyable, it is right now extremely hard to do a thorough library search on a subject: some items are on loan, some are at the book binders, some have been lost, and some are simply not there. Even having found what one wanted to find it is usually extremely time consuming and actually boring to sort the useless from the useful. To enlist the aid of a computerized system in that process is a long overdue step. For instance, the abstracts of articles published in the major refereed journals could be archived on a data base (abstracts in machine-readable form are already available for some journals; also, text scanners could be used). This data base could not only be used for literature search operations, but could actually be a science tool in itself: it could be processed with context understanders, turned into a fact base, and be examined for consistency and inferences in a fully automatic manner. Also, new findings could be checked against such a fact base (Pirenne, 1988).

Carrying the concept a step further, we should begin to build knowledge bases of selected subfields of astronomy. Not only could we, in this manner, combine the expertise of more than one specialist in the field in such a knowledge base, we could also generate new knowledge by processing the combined knowledge.

Of course the ultimate goal would be to extend this approach across different disciplines. This would be a historic step: for the first time we would reverse the trend toward ever increasing specialisation, which was necessitated by the finite capacity of the human brain to handle information. Going back to the "Renaissance" approach of interdisciplinary knowledge processing would immediately yield new insights based on combining pieces of information which are currently separated by discipline boundaries.

14.4.5 The Research Station

The Research Station is in many ways the extrapolation of a data analysis work station. The major difference, however, is that the Research Station would not be a set of peripherals of one particular computer, but rather would be an independent entity, connected to a number of other machines, data bases, libraries, and data taking facilities (locally and remote) through networks, and comprising local compute power. The capabilities of the Research Station should be under the control of an expert system driven user interface, such as the Artificial Reality interface concept (Foley, 1987), which will enable the researcher to efficiently interact with the dynamic environment. In addition, there should be the possibility for the researchers to load their personal knowledge bases into the knowledge server of the station, thus customizing the station to their individual needs and ways of conducting research.

As far as implementation is concerned the Research Station will incorporate many elements which are being used for aircraft cockpit instrumentation and flight simulators. Head-up displays can be used to dynamically display data on top of a scene (i.e. an image), a derivative of the helmet-mounted sight is a much better cursor control device than a joystick. Aural input and voice output will be possible. The goal is to increase the bandwidth of the channel of information between the human user and the computer.

This appears quite elaborate and, at this point, not absolutely necessary. However, as the complexity of the data and the complexity of the operations which we perform on them increase, we will need means of information transfer which are capable of handling this level of complexity. This will allow the "programming" of computers without actually writing software, but through a process of combination of purposeful visual metaphors (Potter, 1988). Similar techniques are already being used to view the results of complex numerical simulations performed on supercomputers, or to visualize the contents of multidimensional bodies of data (Donoho et. al., 1988).

14.5 Improvements of Research Methodology

The most important long-term impact of AI, however, will be the fact that it will change the way we are doing science.

Data acquisition and data analysis, for which we use computers today, are just the beginnings of a process through which the researcher extracts knowledge out of known and new data. Other steps in that process, for instance the building and the refinement of models, the checking of concepts for consistency, or the inferencing from interdisciplinary assertions is still done the way we have been doing it for the last several hundred years. In general, the approach to unveiling the universe so far has been to look harder in order to find evidence for existing hypotheses or indications of new ones. What is being suggested is to try to improve the process of hypothesis generation in order to gain new insights.

Scientists engaged in the pursuit of knowledge in areas which are contained in the term "natural sciences" have at their disposal a set of tools which they use in their investigations. Some of these tools were already identified by the ancient Greek philosophers, others were only developed during and after the period of enlightenment. It is important to realize that these tools change, and that major improvements of the tools always resulted in major new insights. An example is the introduction by Kepler into astronomy of the tool of mathematics (he also was an eminent astrologer, applying to the same data a completely different set of tools; today we like to think that Kepler himself considered mathematics the more important tool).

In astronomy, as in all natural sciences, we are trying to build models of reality with the ultimate objective of coincidence between the model and reality, i.e. finding the "truth". This process suffers from a variety of shortcomings: the fact that the reasoning being used is predominantly inductive; the fact that models might entirely miss important aspects of reality, and even if they do not, there cannot be proof that they do not; not to mention the fact that our model building capabilities were evolved on Earth, so their application to the rest of the universe is questionable at best. However, not only has this process allowed us to gain considerable insights into the universe, it is also the only one available to us.

Models are being constructed from observed facts (data) by applying principles of model building. These are constructs which, when applied to the data, ultimately allow us to make predictions about unobserved (or unobservable) aspects of the model, spatially, in wavelength, and with time. These tools include principles like: constancy, steady change, periodicity, dependence, analogy. They are usually expressed in mathematical notation, and, when combined, quite often lead to complex expressions and operations.

To turn raw data into facts that can be used for model building, they have to be pre-processed. This is known as calibration, parameterization, and classification. A problem in this process is that it usually is not free of underlying assumptions, i.e. a previous model (a hypothesis) exists, which influences the observing strategy, the selection of parameters to determine, and certainly the classification. Conversely, quite often a classification is predetermined by the representation of the data, resulting in an unreasonable model.

14.5.1 Classification

There are two possible ways to classify data: the purely morphological one, (or even physiological one, if the data representation is the dominant element), and one which tries to capture the physics behind the data. Historically, classifications usually start out as morphological/physiological classifications. When refined, they quite often collapse and get replaced by classifications which are more meaningful in terms of physics. This is usually paralleled by a significant change of the underlying model.

These simple considerations illustrate the importance of proper classification in the interrelation between facts and models. The question is how to define "proper", and, once defined, how to implement it. At the same time the classification should be independent of data representation and human physiology, and allow an interpretation in terms of physics.

In principle, we could establish an open ended classification scheme, with the number of dimensions determined by the number of classification parameters, and the bin size governed only by the uncertainty of the data.

Again, this can be done with and without regard of the physics behind the data. If the physics is taken into account, and the underlying model is reasonably correct, such a classification will result in a refinement of the model, i.e. it will be possible to more accurately predict the behaviour of the object. If done without regard to the physics of the object, in other words, if no assumption is made about a previous model, such a classification might yield a totally new model, which covers aspects of the object which were not understood previously. The classification parameters of the second type of classification can be derived from multivariate statistics, or Bayesian classification.

14.5.2 Application of Automatized Cognitive Techniques

Artificial Intelligence is providing us with tools which will be able to help us to establish new classifications and, ultimately, new models. Considerable progress can be expected through proper utilisation of the software products which are already on the market. Currently available expert system shells are providing tools which can be used for classification experiments. Both previously outlined approaches can be followed.

Clearly it is a reasonable first step to try to simulate as closely as possible the classification techniques which have been traditionally used in astronomy, if for no other reason, so as to demonstrate that the system is indeed capable of performing the operation. The problem there is mainly one of knowledge engineering: to capture, fully and consistently, the rules which human experts use to classify data, and to map them into one of the existing expert system shells. An obvious advantage of creating such an automated classification system is the possibility to perform classifications on large data sets without human intervention.

In the long run, however, the second approach will yield the better scientific return, in the sense that it will allow us to create models which otherwise would have been inaccessible. Needless to say, the realisation of such a system is a long-term project and will require substantial preparatory work, both in the area of statistical analysis, and in the utilisation of the expert system tools.

While the automatized search for the optimum data classification will be an enormous advantage, we will, in the long run, proceed to automatize the model building process. As already stated, our model building capability was developed as a result of evolution, and has served us well in the effort of surviving in the Earth-based environment. Research into model building mechanisms will allow us to identify additional model building tools, and to apply them in contexts which our minds are not conditioned to recognize. Combined with the automatized optimum classification, this will make it possible to derive multiple models, based on different aspects of the observed objects. The scientist will then be able to concentrate on finding out which model is closer to the truth.

There are interesting epistemological aspects. Models conceived by humans tend to provide visualization of some sort, even though, as already mentioned, their mathematical representation sometimes is quite complex. When we say we "understand" a model, we do not just mean that we agree with the inferences, but it means that we developed a mental image. This possibility will be removed if we succeed in building models in an automatic manner. If fact, there might be cases when the automatized model building process yields two equally unvisualizable models, both "correct" in the sense that they allow to make correct predictions. Which model is then closer to the truth? One approach to solving the problem is to assume that the more "elegant", more "beautiful" model is the correct(er) one, an assumption which works in traditional model building. Of course it is difficult to quantify "elegance", — maybe the criterion should be how much predicting can be done with as few assumptions as possible.

14.6 Conclusion

The purpose of the above is not to render the scientist superfluous, but to improve the scientific methodology in such a way that it is possible to gain new insights. So far, we basically have been using the computer (and the associated software) to better acquire and more precisely determine our data. We should begin to use computer based techniques to improve the way we convert data into models, in other words to not only process data, but to process concepts.

Obviously there will be instances when the researcher will not understand all the details of all the concept processing which is used to derive a new result. However, even today, we accept computer generated results without checking the processing (the calculations) in detail: we understand "in principle", thus we proceed in our investigation, even though the process sometimes is rather opaque.

Undoubtedly, however, there will be instances when models and solutions derived through these techniques will exceed the ability of a large number of humans to deal with them. It is very hard to attempt to predict the reaction of humans to a scenario in which most of technology and research is dominated by machine intelligence; some attempts have been made (Albrecht, 1988; Stonier, 1988).

In any case, the most important result of astronomical research, the experience of oneself as a seeker, and the contemplation of the results will always be the domain of the researcher.

References

1. Adorf, H.-M. (1989), "Connectionism and neural networks", this volume.
2. Adorf, H.-M. and Murtagh, F. (1988), "Clustering based on neural network processing", *COMPSTAT 1988*, C. Edwards and N.E. Raun (eds.), Physica-Verlag, Heidelberg, 239–244.
3. Albrecht, R. (1988), "On the interrelation between technology and evolution", *Frontiers and Space Conquest*, J. Schneider and M. Leger-Orine (eds.), Kluwer Academic Publishers, Dordrecht, 221–227.
4. Albrecht, R., Adorf, H.-M. and Richmond, A. (1986), "Next generation software techniques", *SPIE Proceedings*, **627**, D. L. Crawford (ed.), 225-230.
5. Donoho, A. W., Donoho, D. L. and Gasko, M. (1988), "MacSpin: dynamic graphics on a desktop computer, *IEEE Computer Graphics and Applications*, July 1988, 51–58.
6. Foley, J. (1987), "Interfaces for advanced computing", *Scientific American*, October 1987, 127–135.
7. Hopfield, J. and Tank, D. (1986), "Computing with neural circuits: a model", *Science*, **223**, 625–633.
8. Miller, G., Johnston, M., Vick, S., Sponsler, J., Lindenmayer, K. (1988), "Knowledge based tools for Hubble Space Telescope planning and scheduling: constraints and strategies". *Proceedings of the 1988 Goddard Conference on*

Space Applications of Artificial Intelligence, NASA Conference Publication 3009, 91-106.

9. Johnston, M. (1987), "An expert system approach to astronomical data analysis", *Proceedings of the 1987 Goddard Conference on Space Applications of Artificial Intelligence*, E. Stolarik, R. Littlefield and D. Beyer (eds.), GSFC Greenbelt, MD.

10. Lippmann, R. (1987), "An introduction to computing with neural nets", *IEEE Acoustics, Speech, and Signal Processing Magazine*, April, 4-22.

11. Pirenne, B. (1988), "The ALEXANDRIA Project", ST-ECF Internal Study Document.

12. Potter, A. (1988), "Direct manipulation interfaces", *AI Expert Journal*, October, 28–35.

13. Rampazzo, R., Heck, A., Murtagh, F. and Albrecht, R. (1988), "Rule based classification of IUE spectra", *Astronomy From Large Data Bases: Scientific Objectives and Methodological Approaches*, F. Murtagh and A. Heck (eds.), European Southern Observatory, Garching bei München, 227–232.

14. Stonier, T. (1988), "Machine intelligence and the long term future of the human species", *AI and Society*, **2**, 133–140.

Glossary, Acronyms and Index

Short Glossary

Terms in italics below also appear as separate entries in this short explanation of important terminology. Further useful terms, defined in this book, may be located in the individual chapters by means of the Index.

Artificial Intelligence: the study of a wide range of disciplines with the common aim of trying to make the bahaviour of computers indistinguishable from the behaviour of humans (cf. the Turing test, described in chapter 2, section 10.1).

Backward Chaining: given values of one or more consequents ("goals"), determine whether they can be derived from (true) antecedents. Backward chaining stops when (true) facts are ascertained. Cf. *forward chaining*.

Blackboard: a global data structure (the metaphor is one of a blackboard in a scientist's office), where partial conclusions and deductions can be entered. Following updating, and when an overall conclusion is attained, information from the *blackboard* may be written into the knowledge base.

Common Lisp: the standard dialect of the *Lisp* language.

Common Lisp Object System (CLOS): an *object oriented* extension to *Common Lisp*. CLOS grew out of (1) Flavors, associated with Symbolics, Inc.; and (2) CommonLoops, developed at Xerox' Palo Alto Research Center (PARC). CLOS has tentatively been endorsed as an ANSI (American National Standards Institute) standard. A good reference on CLOS is S.E. Keene, *Object Oriented Programming in Common Lisp*, Addison-Wesley, Reading, MA, 1988.

Connectionism: another term for the field of artificial *neural networks*. See chapter 13.

Distributed Problem Solving: involves independent processors (or, for that matter, human agents). The mutual cooperation of such autonomous computing agents has to be ensured.

Domain Expert: the person with knowledge or expertise relating to the subject matter of the *expert system*. Most successful expert systems have concentrated on fairly narrow and well-demarcated domains. The *domain expert* provides knowledge indirectly through a *knowledge engineer* or directly to the *expert system* in consultation sessions.

Expert System: in theory, a computer system which emulates a human expert, by asking sufficient questions of various human experts and/or by processing facts stored in databases. In practice, the paradigm of *rule based system* has dominated to date, although other paradigms such as artificial *neural networks* could be more widely used in future.

Firing of a rule: the execution of an *if-then rule*, i.e. deriving all consequences from the antecedents.

Forward Chaining: given a set of implications, often structured as a dependency graph or network (e.g. $a \rightarrow b, b \rightarrow c, d \rightarrow c, b \rightarrow e, d \rightarrow e$, etc.) and given values of some of the antecedents, make inferences about consequents (cf. *if-then rule*).

Frames: are objects as used in *object oriented programming*. (These *objects* have one or more *attributes* or *slots* which can take values. In addition to such *slots*, one may have what are termed *methods* in some systems: these are not simply alphanumeric value(s) but rather calls to procedures to carry out some action. Slots and methods may be inheritable (see *inheritance*).

Fuzzy Set Theory: a theory, developed by L. Zadeh and others, based on sets where membership may be partial. "Crisp" membership is the alternative. A calculus derived from *fuzzy set theory*, and parallelling probability theory in many ways, is *possibility theory*. See chapter 11.

Hypertext: a complex network of information linked by pointers and cross-references, through which the user navigates. The user may read an entry, jump automatically to other entries, find cross-references, and jump back to the first entry.

If-Then Rule: antecedent-consequent pair; cause-effect; supposition-conclusion.

Knowledge Engineer: the computer scientist working on the implementation aspects of the *expert system* (see also *domain expert*). He/she extracts knowledge from the domain expert.

Inheritance: whenever objects (as used in *object oriented programming*; and see also *frames*) are hierarchically structured, it is usual for subordinate objects to have properties (i.e. slot values or methods) of the parent. In the case of more than one parent object with conflicting properties, *multiple inheritance* may be possible. This is where strategies are available for defining inheritance under such circumstances.

Lisp: "list processing" high level programming language, — in fact, the second oldest high level language still in use today. It was developed from 1956 onwards. The large number of dialects which emerged in recent decades have essentially by now given way to the standard *Common Lisp*. Lisp is perhaps the most favoured language for AI applications. The central reference on Lisp is P.H. Winston and B.K.P. Horn, *Lisp*, 3rd Ed., Addison-Wesley, Reading, MA, 1989.

Message Passing: the communication mechanism between objects in *object oriented programming*. This is analogous to procedure calls in *procedural programming* languages. Message passing may be implemented in convenient traditional ways, or may be implemented in hardware when parallel architectures are used.

Natural Language: e.g. Swahili or English as opposed to Lisp or C.

Neural Networks: a new type of computer, made up of a large number of simple processing elements (the "neurons"). The weights (i.e. values stored) on the interconnecting links constitute the "memory" of such a device. See chapter 13.

Object Oriented Programming: programming based on the objects being treated, rather than what actions one is to carry out in specified situations. A paradigmatic case of object oriented programming was in windowing systems. See also *procedural language*.

Possibility Theory: a theory based on *fuzzy set theory* associating possibilites (bounded by zero and one) with events.

Procedural Language: a programming language such as Fortran or C, based on the precise algorithm (sequence of steps) to be followed. A non-procedural language requires the programmer to state *what* is to be accomplished and details of *how* this is to be done are left to the computer. An example is *Prolog*.

Production System: system based on rules (see *rule based system*).

Prolog: a non-procedural programming lanuage, based on mathematical logic ("logic programming"). Developed in France and Britain in the 1970s, it achieved important status when adopted for the Japanese Fifth Generation Project in 1981.

Relational Database: most common current design used by database management systems. Sets of stored data are represented as relations, i.e. as tables or matrices. Alternative designs, or *models*, are hierarchical or network.

Rule Based System: system based on *productions*, i.e. *if-then rules*.

Shell (Expert System Shell): a software system with one or more inference strategies and with the overall structure of the knowledge base, allowing a user to rapidly build an expert system. There are many commercial expert system shells on the market.

Syntactic Pattern Recognition: the analysis of morphological and other patterns using symbols whose mutual relationships are governed by syntactic rules and, possibly, by grammars. *Syntactic pattern recognition* is often counterposed to statistical pattern recognition. See chapter 12.

Acronyms

ADAS	Astronomical Data Analysis System
AI	Artificial Intelligence
AIPS	Astronomical Image Processing System
ANS	Artificial Neural System
ANN	Artificial Neural Network
ANNE	Another Neural Network Emulator
ANSI	American National Standards Institute
AP	Application Programs
ART	Automated Reasoning Tool
ART	Adaptive Resonance Theory
ASM	Administrator of Session Monitors
AXAF	Advanced X–Ray Astrophysics Facility
BDI	Binary Digital Image
BLS	Bidimensional L System
BSB	Brain–State–in–the–Box
CAIL	Conditional Attributed Interaction Lyndenmaier System
CCD	Charge Coupled Device
CDI	Coloured Digital Image
CERGA	Centre d'Etudes et de Recherches Géophysiques et Astronomiques
CLOS	Common Lisp Object System
COMPTEL	Compton Telescope
DAM	Distributed Associative Memory
DAP	Distributed Array Processor
DARPA	Defense Advanced Research Projects Agency
DAS	Data Analysis System
DBMS	Database Management System
DEC	Digital Equipment Corp.
DMF	Data Management Facility
DSP	Digital Signal Processor
DSS	Decision Support System
DT	Decision Table
EA	Expert Assistant
ENA	European Neuroscience Association
ERL	Environmental Reseach Laboratories
ES	Expert System
ESA	European Space Agency

ESO	European Southern Observatory
EST	Expert System Tools
FFT	Fast Fourier Transform
FGS	Fine Guidance Sensor
FITS	Flexible Image Transport System
FOC	Faint Object Camera
Fortran	Formula Translator
FOS	Faint Object Spectrograph
GAME	Generic Applications and Monitors Environment
GCC	GAME (see separate entry) Command Chain
GDS	Guarded Discrete Stochastic
GIF	GAME (see separate entry) Interface
GMD	Gesellschaft für Mathematik und Datenverarbeitung
GRO	Gamma Ray Observatory
GSFC	Goddard Space Flight Center
HD	Henry Draper
HDB	Hierarchical Database
HNC	Hecht–Nielsen Neurocomputer Corporation
HRS	High Resolution Spectrograph
HSP	High Speed Photometer
HST	Hubble Space Telescope
ICNN	International Conference on Neural Networks
IEEE	Institute of Electrical and Electronic Engineers
IL	Interactive Lyndenmaier (Systems)
INNS	International Neural Network Society
I/O	Input/Output
INRIA	Institut National de Recherche en Informatique et en Automatique
ip	Imaging Process (Model)
IQL	Interactive Query Language
IR	infrared
IRAF	Image Reduction and Analysis Facility
IRAS	Infrared Astronomical Satellite
ISO	Infrared Satellite Observatory
IUE	International Ultraviolet Explorer
IUWDS	International Ursigram and World Day Service
JACEE	Japanese–American Cooperative Emulsion Experiment
KB	Knowledge Base
KBS	Knowledge Based System
KEE	Knowledge Engineering Environment
L*STAR	Lockheed Satellite Telemetry Analysis in Real Time
Lconc	Local Concavity
Lisp	Lisp Processor
lr	Linear Representation
LTM	Long Term Memory
ME	Multiple Element
MIDAS	Munich Image Data Analysis System

MK	Morgan–Keenan
MVLT	Multi–Valued Logic Tree
MRSP	Muenster Redshift Project
NASA	National Aeronautics and Space Administration
NDT	Non–deterministic Decision Table
NOAA	National Oceanic and Atmospheric Administration
NOAO	National Optical Astronomy Observatories
NP	Non–Polynomial
NQAM	Normalized quadratic axial moments
NRAO	National Radio Astronomy Observatories
OASIS	Operations and Science Instrument System
OMS	Observatory Monitoring System
PARC	Palo Alto Research Center (Xerox)
PC	Personal Computer
POCC	Payload Operations Control Center
pdf	Probability Density Function
PDP	Parallel Distributed Processing
PE	Processing Element
Pep	Proposal Entry Proposer
QDBMS	Query Language for DBMS (see separate entry)
RC	Return Code
RCS	Rochester Connectionist Simulator
RiVas	ridges and valleys
ROSAT	Röntgen–Satellit
SAIC	Science Applications International Corporation
SC	Slope Code
SE	Session Environment
SERC	Science and Engineering Research Council
SESC	Space Environment Services Center
SI	Sensitivity Index
SIMBAD	Set of Identifications, Measurements and Bibliography for Astronomical Data
SIP	Single Image Pattern
SM	Session Monitor
SMS	Science Mission Specification
SOGS	Science Operations Ground System
SPSS	Science Planning and Scheduling System
STARCAT	Space Telescope Archive and Catologue
STECF	Space Telescope — European Coordinating Facility
STM	Short Term Memory
STScI	Space Telescope Science Institute
STSDAS	Space Telescope Science Data Analysis System
TAC	Time Allocation Committee
TACOS	Time Allocation Committee Operations Support
TDRSS	Tracking and Data Relay Satellite System
TI	Texas Instruments, Inc.
TP	Training Parameters

TSP	Travelling Salesman Problem
UIMS	User Interface Management System
UV	ultraviolet
VLSI	Very Large Scale Integration
VLT	Very Large Telescope
WIMP	Window, Icon, Mouse or Pointer (Interface)
WF/PC	Wide Field/Planetary Camera
YACC	Yet Another Compiler–Compiler

Index

Lecture Notes in Mathematics

Lecture Notes in Physics

L. B. Robinson, University of California, Santa Cruz, CA, USA (Ed.)

Instrumentation for Ground-Based Optical Astronomy Present and Future

The Ninth Santa Cruz Summer Workshop in Astronomy and Astrophysics, July 13 – July 24, 1987, Lick Observatory

1988. 303 figures. XXXIV, 751 pages. (Santa Cruz Summer Workshops in Astronomy and Astrophysics). ISBN 3-540-96730-3

Contents: High Resolution Spectrographs. Chaired by D. Schroeder and S. Vogt. – Low Resolution, Faint Object Spectrographs. Chaired by J. Miller. – Multi-Object Spectrographs. Chaired by H. Ford. – Imaging and Photometry, Image Reconstruction. Chaired by G. Lelièvre. – Data Analysis, Software and Systems. Chaired by S. Grandi. – Detectors I: CCDs. Chaired by J. B. Oke. – Detectors II: Infrared Photo-Cathodes, Photon Counters. Chaired by M. Cullum. – Instrument Control, Data Acquisition. Chaired by P. Jorden. – Telescope Control, Remote Observing. Chaired by J. Geary. – Equipment Available from Industry; Closing Session. Chaired by L. Robinson.

L. Bolc, University of Warsaw, Poland; **M. J. Coombs,** New Mexico State University, Las Cruces, NM, USA (Eds.)

Expert System Applications

With Contributions by numerous experts.

1988. 84 figures, 12 tables. IX, 471 pages. (Symbolic Computation – Artificial Intelligence). ISBN 3-540-18722-7

Contents: Representing Control Knowledge as Abstract Tasks and Metarules. – Controlling Expert Systems. – A Quantitative Approach to Approximate Reasoning in Rule-based Expert Systems. – Structural Analysis of Electronic Circuits in a Deductive System. – Building Expert Systems Based on Simulation Models: An Essay in Methodology. – An Approach to Designing an Expert System Through Knowledge Organization. – Garden Path Errors in Diagnostic Reasoning. – Knowledge Organization and Its Role in Temporal and Causal Signal Understanding: The ALVEN and CAA Projects. – Subject Index.

Springer-Verlag Berlin Heidelberg New York London Paris Tokyo Hong Kong